Interstellar Matters

T0226005

Springer
New York
Berlin
Heidelberg
Hong Kong
London
Milan
Paris
Tokyo

Gerrit L. Verschuur

Interstellar Matters

Essays on Curiosity
and Astronomical Discovery

With 81 Illustrations

Springer

Gerrit L. Verschuur
Bowie, MD 20715
USA

Library of Congress Cataloging-in-Publication Data
Verschuur, Gerrit L., 1937–
 Interstellar matters: essays on curiosity and astronomical
discovery/Gerrit L. Verschuur
    p. cm.
    Bibliography: p.
    Includes index.
    1. Interstellar matter. I. Title.
QB500.V47 1988   88-20016
523.1'12—dc19

ISBN 0-387-40606-9   Printed on acid-free paper.

First softcover printing, 2003.

Printed in the United States of America.

9 8 7 6 5 4 3 2 1   SPIN 10947265

www.springer-ny.com

Springer-Verlag New York Berlin Heidelberg
A member of BertelsmannSpringer Science+Business Media GmbH

Contents

Introduction

Curiosity is the driving force behind the scientific endeavor, and the thrill of discovery often its reward. Our story is specifically about the exercise of curiosity in the pursuit of interstellar matter, which everywhere drifts between the stars of the Milky Way. The discovery of its existence came not as a sudden event that burst upon the world to merit a few days of headlines, but was the result of decades of patient work by hundreds of astronomers.

In order to recognize that interstellar matter exists, a profound change had to occur in the way astronomers viewed the universe. That change began a century ago and took several decades to play itself out. It marked the transition from astronomy to astrophysics. When the transformation was done, humankind was able to open its collective eyes into a new universe, one that contained galaxies, nebulae, supernovae, interstellar matter, pulsars, quasars, and, of course, those notorious black holes.

Interstellar Matters begins with a dramatic saga of discovery which focusses on the life of Edward Barnard, a self-made astronomer who rose from a life of abject poverty during the Civil War to become the world's first great photographer of the heavens. For most of his adult life he struggled to understand the nature of "vacancies" between the stars, which were clearly revealed in his beautiful photographs of the Milky Way. In the second part we find that the notion that there exists matter between the stars is slowly accepted.

This book aims to cross the barriers of tradition. We do not pretend to cover every aspect of every subject considered under the heading of interstellar matter. Instead, a cross section of essays on historical, modern, and philosophical topics are offered and combined with personal views into tricks of the astronomical trade. In the course of these explorations we share, in the third part, a perspective on the thrill of discovery that motivates astronomers. In some cases we will see how discoveries were made. All the stories are set in the context of the manner in which astronomers deal with their findings through the use of metaphor, models, and notions of balance.

The fourth part of the book deals with modern aspects of the study of interstellar matter in such a manner that the reader will gain a sense of just how so much has been learned about the gases between the stars. The new discoveries related to star formation from clouds of interstellar matter are now casting a completely new light on the origin of life. The earliest steps in the process may have occurred in space rather than on earth.

Unavoidably, the personal pronoun he rather than she is often used in this book although we hope that more women will become interested in astronomy in years to come. At present there remains a remarkable unbalance in apparent interest in this subject between men and women. A startling statistic was recently revealed by *Sky and Telescope* magazine when it reported that 94% of its readers were male and only 6% female. We hope that in future the ratio will inexorably change toward greater balance.

This book has benefitted from the help of a number of people to whom I wish to express my special thanks: Donald Osterbrock, Arnold Heiser, David Malin, Pat Shand, James Felten, Charles Lada, Richard Dreiser, Allan Sandage, W. Butler Burton, Eric Duel, Gaylin Laughlin, Elizabeth Lada, Charles Lada, Jeffrey Robbins—my patient and helpful editor—and Joan Schmelz for her unstinting support. I also wish to extend my thanks to Gregory Shelton of the U. S. Naval Observatory library, and the Mary Lea Shane Archives of Lick Observatory.

Sections of a number of chapters have been published elsewhere and I wish to acknowledge *Astronomy Magazine* (chapter 22 and part of chapter 17), and the *Griffith Observer* and the Hughes Aircraft Company for segments of material related to Barnard's dilemma.

Part I   The Genesis of an Idea: Barnard's Dilemma

The discovery of interstellar matter did not burst suddenly upon the astronomical world. It was a tortuous process that played itself out over more than a century, a saga marked by faltering steps as a number of astronomers struggled to confront the notion that perhaps space between the stars was not empty. A key figure in our story is a remarkable astronomer, Edward E. Barnard, who rose from a life of extreme poverty to become the world's greatest photographer of the heavens. He entered the scene at the end of the era when astronomy was largely a question of looking through telescopes. His greatest challenge was to understand the nature of the dark markings revealed in his beautiful photographs of the Milky Way; dark markings, which had long been believed to be vacancies between the stars, "holes in the heavens" through which one could see empty space beyond the stars. Once he began to photograph these markings, Barnard was torn between deciding whether they were, in fact, vacancies or, as some astronomers suggested, indicative of obscuring matter between the stars.

CHAPTER 1

"What Omnipotence!"

Setting the Scene

"So enormous is the number of stars, yet so completely incalculable are they, as to admit of their being joined with the sand upon the sea-shore, as a Figure of speech denoting a numeration which we cannot define."[1] So wrote the Reverend Thomas Milner in 1858 in his book of popular science entitled *The Gallery of Nature—A Pictorial and Descriptive Tour through Creation Illustrative of the Wonders of Astronomy, Physical Geography, and Geology.*[2] Self-explanatory titles were the rage in those days.

Lest his readers be overwhelmed by the size of the starry realm, the Reverend Mr. Milner assured that the Creator was "acquainted minutely with these multitudinous worlds, which immeasurably exceed our utmost estimates." This important point taken care of he proceeded to wax lyrical and gorgeously descriptive on the wonders of astronomy. Most noticeable in the heavens was, "That luminous celestial highway which the Greeks called the Galaxy, and the Romans the Via Lactea, from its whiteness. . . . By some of the pagan philosophers the Via Lactea was regarded as an old disused path of the sun, of which it had got tired, or from which it had been driven, and had left some faint impression of his glorious presence upon it."

Milner's words of clerical awe, written at the time when Darwin was beginning to confront the meaning of evolution, provide a perspective on our path through knowledge about our galaxy, its size, shape, contents and the nature of the dark markings amongst the stars.

A century ago the Milky Way was the name given to the bright band of stars that sweeps across the summer skies in the northern hemisphere. It is even more spectacular in the southern heavens (figure 1.1), where the brightest parts of it are visible high above the horizon. Today it is the name given to the entire galaxy in which we reside. In 1858 the concept of a galaxy was unknown, and the existence of dark regions between the stars barely noticed. These can even be seen with the naked eye as starless bands in the constellation Cygnus of a summer's night, provided you are well away from city lights.

Milner pointed out that the stellar composition of the Milky Way was suspected long before it was proved, and "... its multitudinous host of stars remained a secret till [William] Herschel turned his mighty instrument at Slough upon the silvery belt." It was then that some glimmering of the scope of the Milky Way began to dawn on human consciousness. Herschel apparently estimated that there were no fewer than 258,000 stars in a zone 10° by 2½° across, a gross underestimate by today's standards.

A century later astronomers not only found ways to define the number of stars, as Milner might have wished, but estimated that there were the equivalent of 250 billion suns in the Milky Way.

In the 19th century, as is true today, few people were privileged enough to see the magnificence of the Milky Way from a dark site atop a mountain where atmospheric absorption of starlight is low, and light pollution from distant cities nonexistent. Milner reported on a journey of someone named Brydone to the top of Mt. Etna.

The sky was clear, and the immense vault of the heavens appeared in awful majesty and splendor. The whiteness of the Milky Way was like a pure flame that shot across the heavens, and with the naked eye we could observe clusters of stars that were invisible in the regions below.

"What omnipotence!" was the involuntary exclamation of one Schroeter of Lilienthal, "upon examining a part of the same magnificent girdle." Milner added, "Surely the situation alone is enough to inspire philosophy." Ah yes, surely it is. But it has inspired far more than philosophy, which is defined as a critical study of fundamental beliefs. It has inspired curiosity, expressed as the need to probe beyond belief.

"To measure the distance of the stars, is a task which has baffled the ablest men," the reverend told his readers, "armed with the best instruments for the purpose, and using them with the utmost nicety and perseverance."

How far are the stars? How deep is the Milky Way? Until the early nineteenth century no parallax of one second had been detected and thus Milner knew that the stars had to be more than nineteen billion miles away.[3] Indeed, stars are very far away, trillions of miles, and with the use of new definitions, such as parsecs and light years,[4] astronomers have made the numbers manageable. A century ago the scale of the Milky Way was totally unknown, although astronomers had begun to recognize the existence of nebulae, or island universes which appeared farther away than the stars.

Milner was deeply moved by nature and saw in all its manifestations the power of the creator. Yet he tried to report faithfully what was known about Nature in the mid-nineteenth century with token reference to the power that created it all. The nebulae moved him more than anything else in his narrative. "Here we have firmaments and clusters, insulated in space, each constituting a sidereal family equal to that to which our

FIGURE 1.1. A spectacular wide-angle photograph of a section of the Milky Way toward the Southern Cross, the grouping of four stars that can be seen just to left of center. Like a thumb-print, the Coal Sack, one of the "vacancies" or "holes in the heavens" which so long tantalized astronomers, asserts its presence beside the Southern Cross. The two bright stars near the left-hand side of the region are

α and β Centauri, the pointers to the Cross. The two patches of nebulosity in the lower part of the photograph, well below the Milky Way, are the Large and Small Magellanic Clouds, relatively nearby neighbors of our galaxy. These two galaxies, as well as the Coal Sack, are visible to the naked eye. (European Southern Observatory)

sun belongs." He was not speaking of galaxies as we know them. He believed that the nebulae were swarms of stars and that the sun existed inside one of these. Some of the nebulae later turned out to be what are now called *globular clusters*, closely packed groups of thousands of stars, not unlike Milner's belief, but others were very different beasts.

"One of the most conspicuous of these objects, called the transcendently beautiful queen of the nebulae, and the oldest known, appears below the girdle of Andromeda." Visible to the naked eye in the absence of the moon, it had often been mistaken for a comet and, according to Milner, was first noticed in the tenth century. This is the great galaxy in Andromeda, M31, located 650,000 parsecs (650 kpc)[5] away (figure 1.2).

The most startling shapes that had turned up in the first telescopic views of the nebulae were those of spiral or whirlpool form (figure 1.3). Concerning other shapes Milner wrote: "Two very remarkable nebulae, well known to mariners, being visible to the naked eye, distinguish the southern firmament. They are cloudy masses of light called the Nubecula Major and Minor, greater and lesser cloud, but familiarly referred to as the Magellanic Clouds, after the navigator Magellan, who was one of their first European observers." The larger of these two galaxies, the Large Magellanic Cloud (figure 1.1), recently burst into the public eye when it hosted the first supernova explosion seen with the naked eye since the invention of the telescope.

Milner gradually lead his readers to consider the enormity of what the nebulae appeared to imply. "However it may savour of the gigantesque, it is sufficiently evidenced that an area of the heavens not exceeding 10/1 of the lunar diameter contains a system of stars rivalling those which constitute our firmament, and appearing as only a single faint luminosity to us. Yet there are many areas so occupied."

He moved forward cautiously, revealing that "it follows therefore that our firmament is but one of a series, and probably one of the smaller chambers in the great mansion of the universe."

Poetic, but it was only an approximation. Our firmament, our environment, the Galaxy as we now call it, is but an average player upon the scene. Milner guessed at the truth when he wrote, "We may conclude, however, that as the firmament, which the unaided eye of man surveys, is but a member of a vast family of systems which his assisted vision scans, so that family may be no more than as a drop to the ocean, a grain of sand to the mass of the globe, compared to what lies beyond the bounds of telescopic sight, hid in regions which mortal gaze will never explore or visit."

He was probably right about a visit, but so wrong about what mortal gaze was to explore, especially when it came to be aided by cameras and spectroscopes, and telescopes in the radio, infrared, x-ray and ultraviolet parts of the spectrum.

FIGURE 1.2. A spectacular view of the inner regions of the nearby galaxy M31, in the constellation Andromeda. Visible to the naked eye as a small, fuzzy patch of light, it was first commented upon in the tenth century and for centuries remained the prototype "nebula," even though its nature was not understood until the twentieth century. A satellite trail is visible as a thin white streak across the image. (AURA Inc. National Optical Astronomy Observatories)

FIGURE 1.3. The face-on galaxy NGC 300, photographed by David Malin with the 150-inch telescope in Australia. Its spiral or whirlpool form is characteristic of a large number of nebulae. Before these spiral nebulae were recognized to be distant galaxies there was considerable speculation as to what they might be. (Copyright Anglo-Australian Telescope Board)

The good reverend imagined travelling through space to "take our station at the point which is not the limit of vision" and asked whether we would not see as much again. He knew that the "Divine capability" would not cease to create beyond our limits of vision. "That part of space beyond which the art and genius of man fail to conduct his glance, simply remind us that we are finite beings, and that we have reached the limit which for the present circumscribes our finite powers." But those finite powers would be extended by new inventions, the camera and the spectroscope[6], which when attached to telescopes would allow humankind to explore further and deeper into the voids of space.

The Orion Nebula

One of the first photographs ever obtained of the Orion nebula is shown in figure 1.4.[7] An early description of this object was poetic indeed.

There is food for imagination here and room for questions to satisfy the most speculative minds. To what end this fiery mist? What mighty and mysterious forces are at play? Are these gigantic outer wreaths moving outward or falling in toward the center? Is this the beginning of worlds or is this the end?

But speculative minds are not satisfied by questions alone. They want answers. To understand the nature of the "fiery mist" in Orion, to know whether it was moving this way or that, required the new technology of spectroscopy to get at the answers. With the invention of the spectroscope, which allows the quantitative analysis of the component wavelengths of light (chapter 18), the astronomer had a tool that could be applied to the task of understanding stars, nebulae, and galaxies. Only then would questions be answered that were related to the beginnings or the ends of worlds.

Agnes Clerke, popularizer of astronomy a century ago, described the interpretation of the first spectral lines observed from the Orion nebula. These observations had revealed nebulous "knots" near the Trapezium (the group of stars within the nebula that provide the energy that causes the nebula to shine). She wrote that "this was thought to indicate an advance of central condensation—possibly even the beginning of the long birth process of an orderly evolving system, reserved for future habitation of rational beings. It may be so; the ways of creative power are dark."[8]

This was remarkably insightful for the time, but she was also aware that the picture might not be so simple. "Yet we cannot help remarking that the presence of so many stars fully formed, yet seemingly wrapt up and involved in the prodigious masses of nebulosity filling the portion of the sky, appears to some degree to discount the expectation of stellar development for them." In other words, she could not believe that given so many fully formed stars others might yet be born. Although she didn't

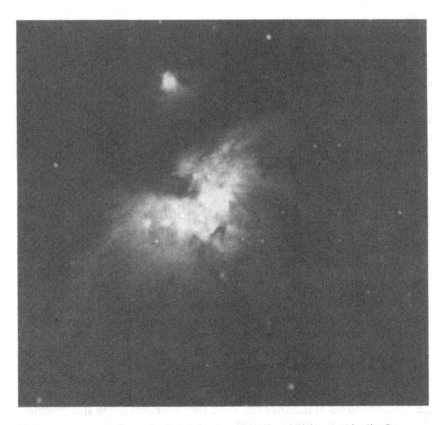

FIGURE 1.4. The Orion nebula as photographed in 1883 by A. Ainslie Common with a one-hour exposure (from *A Popular History of Astronomy*, by Agnes Clerke, 1887, New York: McMillan). In this, one of the first astronomical photographs made using long time exposures the image reveals great detail in the structure of the nebula which consists of incandescent gas.

know it, Ms. Clerke had described a form of interstellar matter, a concept still years in the future.

Figure 1.5 is a modern view of the Orion nebula and dramatically

FIGURE 1.5. The Orion nebula as photographed with the Anglo-Australian telescope by David Malin using a 5-minute exposure. Comparison with Figure 1.4 reveals the extraordinary improvement in photographic capability since Common's day. The four bright Trapezium stars that power the nebula can be seen just below the tongue of dark matter that projects into the butterfly-like structure. Swirling clouds of interstellar matter, both luminous and dark, can be seen throughout the Orion region located about 400 pc from the Sun. (Copyright Anglo-Australian Telescope Board)

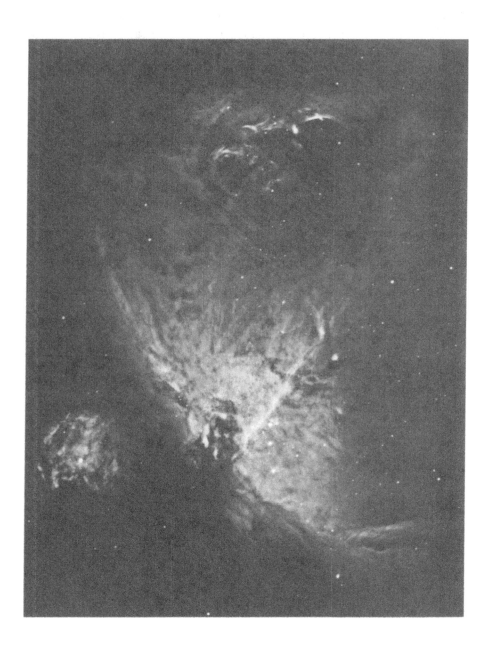

highlights the changes that lay in store for astronomers as they built bigger and better telescopes and learned to see the stars and nebulae more clearly. With the advent of larger telescopes, and improvements in photographic techniques that allowed the images to be captured on film, the presence of dark markings between the stars, some of which can be seen in front of the Orion nebula, would begin to draw attention. And the pioneer in this work, Edward Barnard, was born the year before the Reverend Mr. Milner waxed lyrical upon nature's wonders.

Notes

[1] Milner T. (1858) *The Gallery of Nature. A New Edition.* London: W. and R. Chambers, p. 158.

[2] Ibid.

[3] Parallax is the apparent change in position of a star with respect to other stars as observed from two points in the earth's orbit separated by the sun-earth distance as seen from the star's direction.

[4] The official astronomical unit of distance is a parsec, the distance at which a star would have a parallax of one second of arc. It is 3×10^{18} cm, close to 3.25 light years, or 19 trillion miles. A light year is the distance light travels in one year, or about 6 trillion miles.

[5] The notation kpc, for "kiloparsecs," referring to a thousand parsecs, is commonly used. Thus, M31 is 650 kpc distant.

[6] The principles of the spectroscope, with particular reference to radio astronomy, will be outlined in chapter 17.

[7] The first good photo of Orion was made in 1882 by Henry Draper. Washburn Obs xxv, App. i: 226. Also Comptes Rendus xcii p 261, and Mon. Not. Roy. Astr. Soc., xliii: 233. The photo in figure 1.4 was made by A. Ainslie Common on 28 February 1983, one hour exposure.

[8] Clerke, A.M. (1887) *A Popular History of Astronomy.* New York: Macmillan.

CHAPTER 2

"The Immortal Fire Within"

Childhood

Edward Emerson Barnard was born on 16 December 1857, in Nashville, Tennessee. His father had died two months earlier and his mother had a terrible struggle to support herself and her two sons. When Edward was three years old the Civil War broke out and a few years later Nashville became an important city in the campaign. Tennessee was a Confederate state and Nashville had been occupied by Federal troops in February 1862, an event that was to provide for the survival of the Barnard family in a most unusual manner.

The Cumberland river flows through Nashville into the Mississippi and drains the Cumberland Gap, where intensive fighting took place on September 1862. From 1862 to 1865 Nashville was a major supply base for Federal forces, being a major rail center, and the river was always filled with debris, not all of which was useless. Containers of hard tack (dried biscuits) would drift by and Ed became an excellent swimmer as he dived in to retrieve the flotsam. These biscuits kept the family alive during the war.

The last major battle of the Civil War (between Generals Thomas and Hood) took place in Nashville in December 1864 and the sounds of the bombardment were to haunt Barnard for the rest of his life.

This man became an heroic example of how far an individual can go if he sets out in a dedicated manner to pursue his goals. Despite an almost entire lack of formal schooling, he rose to accomplishments that even the most educated seldom achieve. He travelled a long way in body and spirit from this childhood of poverty and the sounds of gunfire to use the greatest telescopes at the foremost observatories in the world.

As we pursue his path through life we ask how it came to be that a person born of so humble a beginning, of so poor a situation, could manage, with only two months of formal schooling in his teen and preteen years, to walk with the world's greatest astronomers. In answer, Alfred E. Howell, a resident of Nashville who knew Barnard in the early years, wrote, "It was the immortal fire within himself that made him great."[1] This fire smoulders within us all but is too often doused in early childhood or extinguished in one's teens. We can only hope that when we confront it in each other we may have the wisdom to fan the flame into a blaze, that teachers, guides, and mentors will have the patience and love to encourage their students' curiosity and search for truth.

Barnard survived an early attack of cholera and from his ninth year (figure 2.1) worked in order to support the family.

His mother was a woman of character, and did not lose her taste for culture in the struggle with poverty. The name, Emerson, given to the young son in honor of our American philosopher, was evidence of this. His mother had a taste for art, also, and partially supported herself by modelling wax flowers. That she had impressed on her son character and self-reliance was indicated by her statement that the lad, when less than nine years old, could be depended upon to do a task in which many other boys had failed.[2]

This task was to continually point a ponderous enlarging camera on the roof of Van Stavoren's Photographic Gallery at the sun. Barnard was a human substitute for a driving clock and where other boys had gone to sleep at the task, he succeeded. He worked at the studio, in various capacities, for seventeen years. During that time he developed the tremendous patience and concentration that would turn him into a famous astronomer.

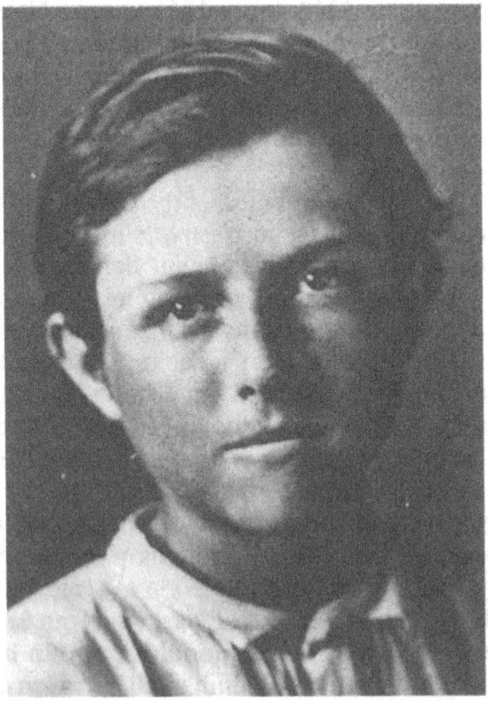

FIGURE 2.1. Ed Barnard around the time he began his labors at van Stavoren's Photographic Gallery at age nine (1866). In this job he worked diligently to keep the giant enlarger, "Jupiter" (Figure 3.2) tracking the sun. (Vanderbilt University Library Archives)

The giant enlarging camera was known as "Jupiter" (figure 2.2). A man sat inside the device (no hand held automatic this). In order to enlarge an image from the small negative a great deal of light was required. How did one obtain an intense beam of light in days of insensitive film and no electrical power? One used sunlight, of course. The time exposures were considerable. Sometimes Van Stavoren painted from the image (an early form of photorealism) and had to have it fixed in position for great lengths of time.

If one set up the enlarging camera to point at the sun, the sun's motion soon moved the beam of light out of alignment and the image moved

FIGURE 2.2. The awesome "Jupiter" enlarging camera, which Ed Barnard had to keep pointed at the sun by continuous adjusting the wheels at the side. His experience at keeping this contraption tracking accurately was to serve him well when he became an astronomical photographer. van Stavoren, proprietor of the Photographic Gallery, is on the left and J.W. Braid is inside the camera while Barnard is sitting on his chair. This photo must have been taken around 1877, when Barnard was twenty years old and as yet unaware of his astronomical bent. (Vanderbilt University Library Archives)

across the focal plane, the screen at the back of the camera. That would not do, so the device had to track the sun. Nowadays one can go to the neighborhood hobby store and buy a telescope drive that can perform the task; but not in 1870. The function of tracking the sun had to be performed by cheap labor—little boys. This was cold work in the winter and in the summer the warm sun had made other boys fall asleep at the job. This posed a considerable hazard since the focussed sunlight could start a fire in the wooden structure if not watched carefully. It was Mrs. Barnard who pressed Van Stavoren to hire her son for the job at which all the other neighborhood kids had failed.

It was at the task of turning the wheels to keep the mighty Jupiter tracking the sun that Barnard became adept. Thus, at the age of nine, his training to become the world's greatest photographer of the heavens had begun. No one, least of all Edward himself, could have imagined where his destiny lay.

The Curious Observer Emerges

A simple pinhole had been rigged atop the camera so that an image of the sun was projected on a screen. Young Ed had to drive the camera—that is, turn large wheels (figure 2.2)—to keep the image on the same spot for hours on end. The camera was really a solar telescope. Despite the fact that the tedium of this task had overwhelmed all who had tried before, driven by the necessity of supporting his family, the boy performed well and did not fall asleep at the job.

On the contrary, he turned the task into an opportunity, exhibiting a healthy dose of curiosity in the process. He began to notice interesting things; for example, the shadow of "Jupiter" seemed to move about in an odd way. He made a habit of marking where the shadow fell each day when the noon bell rang at the nearby Catholic church. The mark moved and could not be relied upon to indicate noon. In the middle of February the sun and bell kept together for a couple of weeks, but by the end of March the shadow had shifted so that noon, as defined by the sun's highest excursion in the sky, arrived after the bell. Then the sun's motion slowed down in May and speeded up again around the middle of June. The fastest change from day to day came around the end of the year. These observations, which very few people have ever experienced directly, meant a great deal to Barnard.

"As I could not suspect the church bell of being wrong, I decided there must be something wrong with the sun."[3] And so it was. The sun was moving in a peculiar way with respect to the earth. This happens as the result of a complicated relationship between the changing orbital velocity of the earth and the tilt of the earth's equator to the plane of its orbit, giving rise to what is known as the *equation of time*. Edward Barnard had made his first astronomical discovery, but didn't know what it meant.

His First Telescope

In 1870 J.W. Braid from New York City was employed as operator at the Photographic Gallery. Barnard, who was now thirteen, still patiently filled the role of guide-motor. During the long hours on the job he became interested in the properties of lenses and told Braid that he wished he could get some and fix them up to look at the moon and planets. The youngster had already become addicted to viewing the heavens.

"I can recall when I was a small child," he was to write in the *Christian Advocate* in 1907, "lying out in the open air in an old wagon bed, flat upon my back on pleasant summer nights, watching the stars. I soon knew the stars thoroughly as the seasons came and went."[4] He did not yet know the names of the stars or constellations because there was no one to teach him, yet he came to treat them as friends who would return to visit every year. "I soon saw that a few of them changed their places among the others, but I did not know that they were the planets."

Braid was soon able to help this unusually observant boy in an fortuitous manner. "One day when I was in the shop of Mr. Charles Schott I noticed an old ship's spyglass hanging on the wall," he reported.[5] "It had no lenses. He had bought it for old brass, and intended to cut it up. I made an offer to buy it, and when I told him I wanted it for Ed Barnard he let me have it for two dollars."

In due course he and Barnard sent to Queen and Co. of Philadelphia for an objective lens of 32-inch focal length. "We made an eyepiece out of the wreck of an old microscope, which gave a power of about thirty-eight. A simple altazimuth mounting was made. (Braid was an accomplished instrument maker.) Barnard had a good tripod which had been a surveyor's instrument stand."[6]

The scene was set. The teenager, together with his mentor, went out into the night and Barnard turned the telescope on the planet Venus, which showed as a beautiful crescent. "He was delighted," Braid said. Barnard then viewed the moon, looked at Jupiter, saw its four moons, and observed several sets of double stars, such as the one in Ursa Major. "The simple telescope gave Barnard more pleasure than anything else in his whole life," wrote Braid. The feeling Ed had upon looking through his telescope that night produced in him a thrill which determined his life's work. Virtually every clear night for the rest of his life he pursued the joy of that first occasion; looking at the heavens through a telescope to see new wonders, some of which had never been seen before by anyone else.

"In 1871 Van Stavoren failed in business, and his property had to be sold to liquidate his indebtedness to the Peoples Bank."[7] By then Barnard had built a larger 3-inch telescope which apparently did not function as well as the original 2¼-inch objective lens. When Rodney Poole purchased the photographic gallery equipment he kept the employees and continued the business. Jupiter was dismantled and Barnard moved his telescope onto the vacant platform on the roof which he proceeded to use as his observatory where he began to entertain friends and visitors with views of the heavenly bodies.

Poole's Photographic Gallery developed a specialization in outdoor photography. Braid became chief photographer and Barnard his assistant and they took some of the earliest pictures of the Vanderbilt grounds and other scenes in Nashville. On their outdoor excursions they used Poole's View Wagon (figure 2.3) which was a darkroom on wheels pulled from place to place attached to a one-horse cart.

Alfred E. Howell, a friend of Barnard's, later wrote that "Many [was] the time I have noted his hollow eyes and faded cheeks and wist not that he was to be world famous from his vigils of the night, after an all-day's work at Poole's photographic gallery, his hands still stained with the chemicals."[8] In extolling Barnard's success Howell wrote, "I do not think that we have made too much of our heroes of the stadium, but too little of our seers of the classroom."[9] This sentiment continues to ring true and is still ignored. "He was a member of my Aunt's (Mrs. Fanny D. Nelson's) Sunday school class—he and J. W. Braid and Peter Calvert, whose sister he married."[10] Mrs. Nelson and the Sunday school connection were to become profoundly important influences in Barnard's life, the lady being, "a woman of extraordinary, even brilliant learning."

At this time Braid himself played an important role in the another matter, the development of the telephone, for which Barnard later wanted to make sure Braid received recognition. His mentor had achieved a phone link over seventy miles whereas Alexander Graham Bell had only managed one or two miles at the time.

In his spare time Braid ground lenses in the gallery and described the eager teenager's travels to find a "real" telescope. "Barnard's first ride on a railroad took him to Pittsburgh in search of John A. Brashear who was making telescopes in a small way.[11] On arriving at Brashear's home Barnard was too timid to knock at the door, and paced back and forth till he was weary before he dared make himself and his errand known. Needless to say he had a hearty reception" and an enduring friendship was born.[12]

In 1875 Poole's Art Gallery hired a new person, P. R. Calvert, as retoucher and colorist at the gallery. He had just arrived from England and had some art experience. Barnard liked Calvert's drawings (figure 2.4) and he shared with the newcomer his own drawings of Mars, Saturn, and Jupiter as seen through the telescope. The two became close friends. Ed Barnard was now eighteen years old and had become the printer at Poole's gallery, while Calvert was the retoucher and Braid the photographer. These three worked together every day for the next five years.

"Barnard was never, like other young men, bent on pleasure and frivolous pursuits; necessity and responsibility made him serious, for from

FIGURE 2.3. Poole's View Wagon, a portable darkroom. Barnard, seen here, was J.W. Braid's assistant when the two went out on outdoor photography assignments. (Vanderbilt University Library Archives)

the age of about twelve years he was the main support of his invalid mother who was unable to walk and with beclouded mind, her one joy in life was her devoted boy...."[13] We can forgive Mr. Calvert for not realizing that the pleasures experienced by Barnard were of a different sort, certainly unlike those Calvert believed other young men to pursue. But then other young men had not enjoyed the thrill of discovery, the peak experience, which was Barnard's joy when he first used his telescope to study the moon and planets.

FIGURE 2.4. Barnard with his 5-inch telescope, with which he discovered his first comet in 1881 and the second in 1882. This pen sketch was made by his brother-in-law, P.R. Calvert who worked with Barnard at the photo gallery. (Vanderbilt University Library Archives)

A Gift from the Gods

Although he was always watching the heavens from the roof of Poole's Photographic Gallery, Barnard knew nothing about astronomy. He recognized patterns of stars but knew none of their names nor those of the constellations. There was no one to teach him in Nashville. What little schooling he had was from his mother and that did not include astronomy. Although she was often in poor health, she had taught him to read and write and instilled in him a spirit of independence and self reliance. Yet, no matter how much he wanted to know about astronomy and the names of the stars, there weren't even any books to turn to for information. But then fate intervened.

One day a young man Barnard had befriended and had on occasion lent money, a born thief constantly in trouble with the law, it turned out, showed up on his doorstep. The borrowed money had never been repaid and now Ed's patience ran out when this man came to borrow more money. But the thief was prepared. He had brought collateral.

"On this night I was in no mood to be gracious for he had come to borrow money from me which I knew from experience would never come back. As security for the return of the money he had brought a large book. This I refused to look at. . . ."[14] Finally, to get rid of him, Barnard gave the man the $2 he wanted to "borrow."

After the thief left Barnard, much to his inexpressible delight, discovered that the book was about astronomy. It had the enchanting title *The Sidereal Heavens and other subjects connected with astronomy, as illustrative of the character of the deity, and of an infinity of worlds.* Published in 1840, the author was Thomas Dick, a Scottish philosopher and "divine," who also wrote *The Christian Philosopher.*

Excitedly he skimmed its pages which included star charts that gave the names of the stars with which he was so familiar. "It was but a few minutes until I was comparing these with the sky from my open window, and in less than an hour had learned the names of my old friends . . . ," Vega and Altair and the other stars and constellation he had known since childhood.

"It is to be hoped that my sins may be forgiven me for never having sought out the rightful owner of that book in all these long years," he later wrote.[15]

Thomas Dick Tells of "Holes in the Heavens"

Barnard absorbed the the words of the Scottish cleric. "The delicious old book turned the attention of the boy to the heavens and thus his life work was determined beyond recall."[16] And it was what this book had to say about "holes in the heavens" that was to play a profound role in determining when astronomers the world over would come to understand the nature of the dark markings between the stars.

Thomas Dick wrote about the first telescopic observation of the Orion nebula (figure 1.4), which dated back to 1656 when Huygens drew attention to "one phenomenon among fixed stars worthy of mention," which, so far as his knowledge went, had "hitherto been noticed by no one." He had seen the nebula in the sword of Orion where the "space around there seemed far brighter than the rest of the heavens."[17] Sir William Herschel, according to Dick, believed that when astronomical techniques evolved sufficiently astronomers would find that the Orion nebula "may well outvie our Milky Way in grandeur." This did not come to pass, because the Orion Nebula is but a small region lost in the splendor of the Milky Way itself.

Dark regions among the stars were also mentioned by Dick, who again drew upon the words of Herschel when he wrote that, "in certain directions the spaces . . . were generally deprived of their stars, so as often to afford many fields without a single star in it. . . ."[18] These, Herschel thought, were "great cavities or vacancies" formed "by the retreat of the stars towards the various centers which attract them. . . ."[19]

The Reverend Mr. Dick drew liberally from Herschel's monumental work entitled *The Construction of the Heavens* and in which he had described how he believed the luminous regions (e.g., the Orion nebula) and the dark regions fitted into an overall picture.

So here is young Ed, starved of astronomical knowledge, reading Thomas Dick in *The Sidereal Heavens* and soaking up every word. The book Barnard had obtained for $2 communicated Herschel's work in words meant for the unlearned. "It is an opinion now very generally entertained," wrote Dick, "that the self-luminous matter to which we refer is the chaotic materials out of which new suns or worlds may be formed. . . . Nor do we conceive that this hypothesis is inconsistent with what we know of the attributes and operations of the Almighty. . . . All that we require on this point is some more direct and decisive proofs of the validity of the hypothesis we are now considering."[20] The extraordinary fact is that many of the self-luminous nebulae are indeed regions of star formation, as modern research would inevitably reveal.

Dick, in his gentle attempts to popularize astronomy within the context of the religious tenor of his times, threaded a delicate path through the realm of fact and hypothesis so as not to offend righteous believers. He reported that Herschel considered that the nebulosity might develop and change into stars, the stars develop into dense clusters which would, in turn, attract more matter (stars) to them. Thus, according to Herschel, would be formed voids or "holes in the heavens"[21] (figure 1.1), regions drained empty of stars by the gravitational attraction of large concentrations of matter, the visible nebulae. In formulating so detailed a picture, Herschel was running counter to Biblical accounts of creation, but at least he had suggested an explanation for the dark markings.

Barnard was to carry the word of Herschel's authority, and this image of the holes in the heavens, with him for the rest of his life.

Sir William

Herschel is one of a series of unusual gentlemen who stamp their influence on our story. Born in 1738 in Hanover, Germany, he moved to London in search of work as an oboist when he was twenty. The son of a bandsman, he was himself a born musician, having played the violin when four years old and was a member of the band of the Hanoverian Guards at fourteen.[22] But in London work was scarce, so he travelled around England and survived by playing in public concerts, conducting, composing, and giving lessons. In 1766 he was appointed organist at the Octagon Chapel in Bath, where he settled and made a respectable living which allowed him to live as a "gentleman." He had had no formal training in science and as a hobby began to learn astronomy from books.

It was his interest in music that led to an interest in mathematics which in turn evolved into the study of astronomy. He bought his first book on astronomy when he was thirty-five years old (1773), a book written by the amateur astronomer James Ferguson which bore the title *Astronomy explained upon Sir Isaac Newton's principles, and made easy to those who have not studied mathematics.*[23] It appears that the fear of mathematics kept potential students away from astronomy even two centuries ago, a syndrome that continues to cause universities to offer introductory courses for nonscience majors with the promise that math will not be mentioned.

Herschel began to construct his first telescope soon after reading Ferguson's book. He was renowned for exhibiting extraordinary enthusiasm for whatever interested him. Soon he was learning more about optics, mechanics, electricity, and astronomy, even as he was carrying out his trade of musician and music teacher.

His discovery of the planet Uranus as the result of long and patient surveys of the sky with his telescope brought him instant and enormous fame.[24] The year was 1781. At the beginning of the year he was still a musician whose life was marked by struggle and obscurity. By the end of the following year he was the King's Astronomer, soon to become Sir William, Fellow of the Royal Society, winner of its prestigious Copley Medal. He would be world famous for having been the first person ever to discover a planet. (The five visible planets had been known since prehistory and no one's name is associated with their discovery.)

King George was so impressed with Herschel that he granted the budding astronomer a salary. Herschel moved to Slough near Windsor Castle where, together with his sister Caroline who was also given a stipend, he was able to pursue astronomy as a career until he died at age eighty-four.

That Herschel was a genius few historians doubt. His astronomical labors were prodigious and his legacy reached far into the future. Herschel's view of the universe, however, was heavily influenced by what he had read in Ferguson's book. While most other books on the subject dealt only with the solar system, Ferguson considered the entire starry realm and his universe was a three-dimensional wonder, bordering on modern concepts of vastness.

What an august! What an amazing conception, if human imagination can conceive it, does this give the works of the Creator. Thousands upon thousands of Suns, multiplied without end, and ranged all around us, at immense distances from each other, attended by ten thousand times ten thousand worlds, all in rapid motion, yet calm, regular, and harmonious, invariably keeping the paths prescribed for them; and these worlds people by myriads of intelligent beings, formed for endless progression in perfection and felicity![25]

Ferguson also dwelt upon descriptions of the nebulae which he called "lucid spots" or "cloudy stars." The Orion nebula was "the most remarkable of all the cloudy stars," he said.

It looks like a gap in the sky, through which one might see (as it were) part of a much brighter region. Although most of these spaces [referring to the nebulae] are but a few minutes of a degree in breadth, yet, since they are among the fixed Stars, they must be spaces larger than what is occupied by our solar system....[26]

Of considerable significance to our story is that here the concept of gaps or spaces between the stars was clearly expressed. This notion formed an impression on Herschel, who later would extend the concept which caused Thomas Dick to write about it so that, in turn, an impressionable youngster aching to learn about the heavens would find the description in a stolen astronomy book. Such are the connections that determine our lives.

Barnard Meets a Professional

The thrill of reading Dick's book made Barnard even more anxious for a larger telescope and he heard of one owned by a family named Acklen living at Belmont. How best to obtain this telescope? Perhaps he could write and ask them to loan it to him. But he feared he could not write well enough for so important a task. Thus he convinced his colleague and friend, Peter Calvert, to write the letter on his behalf, which Calvert agreed to do, at a small price, one which would forever change the course of Barnard's life. In exchange for the services of the scribe, Barnard agreed to accompany Calvert to Sunday school. There he met the remarkable Mrs. Anson Nelson, "a refined and cultured lady, a beautiful character, who became one of his best friends."[27] Her husband was an influential man "who met and knew every man of importance in Nashville." This turned out to be the connection that really made possible Barnard's astronomical career. Meanwhile, the desired telescope was sent by express for Barnard's inspection but he found it not nearly as good as he had expected it to be so he returned it with his thanks.

In 1877 the American Association for the Advancement of Science (AAAS) held a meeting in Nashville. Simon Newcomb, Director of the U. S. Naval Observatory in Washington D.C., was its president. The meeting was held at the capitol building in the center of Nashville and Nelson took Barnard, now twenty years old, to meet Newcomb. Many young people have had the experience of meeting for the first time a famous (or even not-so-famous) professional working in the field in which they are interested. Such meetings can inspire or shatter, excite or depress, depending only on random factors such as of luck and ill-formed judgement. The young person may enter such a meeting believing that the authority Figure is superhuman if not quasi-omnipotent. So Barnard believed and his great opportunity left him in tears behind one of the big columns of the capitol building.

Newcomb met the young photographer so inspired by astronomy, who had risen from a life of poverty, and now wished to know what he might do that might be useful in astronomy. Newcomb doubted that this young man could make good. He judged that eager Ed would be able to do little more than amuse himself with his telescope. But then Newcomb did not believe telescopes were of much use in any case, being himself a theoretician. Furthermore, innate snobbery, which so often manifests between professional and amateur in any field, intruded during the meeting. Somewhat brusquely Newcomb advised Barnard to put away his telescope and study mathematics if he wanted to achieve anything at all. Telescopes were of little use, he said. (There are theoretical astrophysicists today who think as Newcomb did.)

Barnard was profoundly depressed but not destroyed by the encounter, as so easily he could have been. It was beyond his means to study mathematics in a formal way so, despite his initial reaction of devastation and tears, he hired a private coach to help him learn. But before the meeting ended Newcomb suggested that if Barnard insisted in wanting to use a telescope he could search for comets, a subject about which Newcomb knew virtually nothing, although he later confessed that he told Barnard what little he did know. Despite the subsequent tears, the comet idea was to change Barnard's life.

Fifteen years later Newcomb recalled this meeting with embarrassment. "It is now rather humiliating that I did not inquire more thoroughly into the case. It would have taken more prescience than I was gifted with to expect that I should live to see the bashful youth awarded the gold medal of the Royal Astronomical Society for his work."[28] Herein lies the moral of the story. When you are young do not hesitate to approach the high and the mighty (even if that requires more courage than you think you have) and when you are old and wise do not hesitate to provide advice and encouragement, no matter whether the person seeking your guidance appears to be hopelessly ill-prepared, or perhaps even a hick from the outback!

Oddly enough, Newcomb received the Gold Medal of the Royal Astronomical Society in the year that Barnard first approached him. Barnard received it in 1892. Barnard was to follow in Newcomb's footsteps again when he was given the Bruce Gold Medal of the Astronomical Society of the Pacific in 1918, a medal Newcomb had received in 1898.

The Comet Hunter is Born

Barnard was inspired by attending the AAAS meeting at which he savored a little of the real world of science. Then he began to search for comets. On 12 May, 1881 he found "a very faint comet in the field with the star α Pegasi." He saw it again the next night, but by the fourteenth it was gone and no one else saw it. Thus his discovery could not be accepted, but in view of his later reputation as an observer his discovery need not be doubted. (As far as his comet discoveries are concerned, during the period 1885 to 1888 the scorecard was to read: Barnard–7; Brooks of Phelps, New York–8; rest of the world–4.)

Although Barnard had discovered a comet he was unable to measure its position accurately. Expressing his concern led to his meeting Olin H. Landreth, who in 1879 had taken charge of the Vanderbilt School of Engineering, and who knew about position measurements (geodesy) from his engineering background. Landreth had worked at the Dudley Observatory in Albany, New York. He was immediately impressed by Barnard's "boundless enthusiasm, by the remarkable store of astronomical information he had already accumulated, and by his wonderfully keen mind, and his sharp eye."[29] Barnard was put in touch with the International Astronomical Bureau of Harvard Observatory and so began to become aware of what was happening in the astronomical world. But, as history was to reveal, communication between astronomers as among others of the time, was not at the level to which we have grown accustomed today.

Landreth showed Barnard how to construct a micrometer for his telescope so he could begin to make accurate measurements of a comet's position in the sky and thus obtain the data needed by the International Astronomical Bureau. By this time the university had a small observatory which had the typical equipment of an educational institution. Because astronomers trained to operate such facilities were scarce the university had not employed anyone to use the idle equipment. Landreth saw this as an opportunity to both train and at the same time hire someone who was of "far more than average promise." But to facilitate the hiring of the untrained Barnard was no easy task, for academics expect to hire their own to fill academic positions. The young man obviously did not qualify on this basis. Nevertheless, Landreth went to the chancellor to suggest he offer Barnard a position.

"Being unprecedented and quite in conflict with fundamental educational principles—or at least practices—the suggestion was not at first kindly received by Chancellor Garland."[30] The president of the University, Bishop H.N. McTyeire, was, however, more understanding, and ". . . wisely forseeing Barnard's capabilities, proved to be sufficiently radical, or at least progressive, to approve the plan and convert the chancellor to its adoption."

When the bishop first asked Barnard to visit him it required the consent of Rodney Poole, at the Photographic Gallery, who had to give permission for the young man to take time off work. From the start Poole expressed willingness to release Barnard from his services should he secure a place at the university.

In 1883 Barnard was appointed Assistant Astronomer of the Vanderbilt Observatory, although he was assistant to no one. The salary was $300 per year, a decrease in salary from his position at Poole's, with which he had to support his young wife and invalid mother. The university also provided him with a house on campus and being in charge of the 6-inch telescope made it worthwhile for him. He also became a special student in the academic department of the university and after four years of study, at the age of thirty, obtained a Bachelor of Science degree. By then his fame as a comet finder was also established.

The Comet House

In January 1881 Barnard married Rhoda, sister of the Calvert brothers with whom he worked at the photo studio. That year was in the middle of a period of hard times, nationwide, and the photo business was not doing well. Yet the groom managed to pay for their first house in a quite exceptional manner. In 1876 he had bought a 5-inch telescope which cost $380, eight months wages for him. Considering that he also supported his mother it is remarkable that he had this much money. Yet, by 1881, he managed to borrow money to buy a piece of land and build a small frame house on it. The first payment came due in September and he found the money thanks to a wanderer among the planets. In that month he officially discovered his first comet. It earned him $200 from H. H. Warner, a patent medicine manufacturer in Rochester, maker of Warner's "Safe Care," who offered this princely sum to anyone who spotted a previously unknown comet.

It was Barnard's optimism about his chances to win some of these awards that made him build the house in the first place. As he was later to report, "when the first note came due a faint comet was discovered wandering along the outskirts of creation, and the money went to meet the payments. The faithful comet, like the goose that laid the golden egg, conveniently timed its appearance to coincide with the advent of those dreadful notes. And thus it came about that this house was built entirely of comets."

Five times in all Barnard received this award ... and it meant to his family the possibility of owning their modest home. That dwelling is still [in 1923] known in Nashville as 'Comet House.' Few, indeed, are the astronomers whose keen eyesight and extraordinary diligence in the quest for celestial discovery have literally provided them with a roof to sleep under. It was very little, however, that he slept under that roof when the sky was clear.[31]

Barnard had an extraordinary dream, soon after discovering his second comet.

My thoughts must have run strongly on comets during that time, for one night when thoroughly worn I set my alarm clock and lay down for a short sleep. Possibly it was the noise of the clock that set my wits to work or perhaps it was the presence of that wonderful comet which was then [October 1882] gracing the morning skies, or perhaps, it was the worry over the mortgage and the hopes of finding another comet or two to wipe it out. Whatever the cause, I had a most wonderful dream. I thought I was looking at the sky which was filled with comets, long-tailed and short-tailed and with no tails at all. It was a marvellous sight, and I just began to gather the crop when the alarm clock went off and the blessed vision of comets vanished."[32]

He woke, took his telescope outside, and began to scan the heavens. To his utter surprise he found a cometary-looking object where none was known.

Looking more carefully I saw several others in the field of view. Moving the telescope about I found there must have been 10 or 15 comets at this point in space of a few degrees. Before dawn killed them out I located [measured the position of] six or eight of them.

He sent a telegram to report his discovery but it was not forwarded in the appropriate way so he never got credit for the discovery. The group of comets was soon confirmed in other countries. What he had seen was a comet after it broke into several pieces by the gravitational pull of the sun. "The association of this dream with the reality has always seemed a strange thing to me."

In all, Barnard discovered seven comets between 1881 and 1887, and seventeen during his lifetime, only one of which he found with the aid of photography. Comet House became famous, as did Vanderbilt University, because Barnard's success as resident astronomer also reflected on the university.

From the start of his career as an astronomer, Barnard's social life hinged around the weather. When invited to an evening function he would accept, under the condition that he show up only if it was rainy or cloudy.

Barnard was an enthusiastic addict of the heavens, a syndrome that is more openly evident in amateur astronomers than many professionals.

He was sometimes referred to as "Enthusiastically Energetic Barnard," to fit his initials. His enthusiasm drove him to work long nights which, combined with continuing difficulty in getting to sleep during the day, inevitably exhausted him. Sleep he regarded as a waste of valuable time.

Notes

[1] Howell, A.E. (1928) *Journal of the Tennessee Academy of Science* , Vol III, No. 1. Edward Emerson Barnard Memorial Number: 8.

[2] Frost, E.B. (1923) "Edward Emerson Barnard." *Astrophys. J.*, 58:1

[3] Hardie, R. (1964) "The Early Life of E. E. Barnard." Leaflet Number 415, *Astron. Soc. Pac.*

[4] Ibid.

[5] Braid, J.W. (1928) "First employment of Barnard: His first telescope." *Journal of the Tennessee Academy of Science* , Vol III, No. 1. Edward Emerson Barnard Memorial Number: 8.

[6] Ibid.

[7] Ibid.

[8] Howell, op. cit.

[9] Ibid.

[10] Ibid.

[11] Brashear was to become one of the the best telescope makers in the United States if not the world.

[12] Parkhurst, J.A. (1923) *Journal. Roy. Astron. Soc. Canada* . Vol. XVII, p. 97.

[13] Ibid.

[14] Hardie, op. cit.

[15] Hardie, op. cit.

[16] Parkhurst, op. cit.

[17] From *Modern Astronomy* by Hecter Macpherson, Oxford Univ. Press, 1928.

[18] Hoskin M.A. (1963) *William Herschel and the Construction of the Heavens.* New York: Norton and Co: p. 80.

[19] Lubbock, C.A. (1933) *Herschel Chronicle* . Cambridge Univ. Press, p. 194.

[20] From *The Sidereal Heavens* by T. Dick, quoted Michael A. Hoskin, *William Herschel and the Construction of the Heavens*, p. 132.

[21] From *Modern Astronomy* by Hecter Macpherson, Oxford Univ. Press, 1928. p. 132.

[22] *Sir William Herschel* by Robert G. Aitken. Leaflet No. 156. *Astron. Soc. Pac.* 1942.

[23] The book was first privately published in 1756. The fourth edition appeared in 1770 and may have been the one Herschel read.

[24] For a while it was suggested that the new planet be named Herschel. Imagine what terrible problems astrologers would now confront if horoscopes had to describe your fate because Herschel was in Sagittarius, or Herschel stood in opposition to Venus at the moment of your birth. A touch of the glamor associated with the name of an ancient god such as Uranus would have been lost.

[25] Ferguson, J. (1778) *Astronomy explained upon Sir Isaac Newton's Principles* quoted in J.A. Bennet, "Herschel's Scientific Apprenticeship and the Discovery

of Uranus" in *Uranus and the Outer Planets*, edited by Garry Hunt, Cambridge Univ. Press, 1982.

[26] Ibid.

[27] Calvert, P.R. (1928) "Reminiscences of Barnard." *Journal of the Tennessee Academy of Science*, Vol III, No. 1. Edward Emerson Barnard Memorial Number: 11.

[28] Ibid.

[29] Landreth, O.H. (1928) "Barnard at Vanderbilt University Observatory." *Journal of the Tennessee Academy of Science*, Vol III, No. 1. Edward Emerson Barnard Memorial Number: 15.

[30] Ibid.

[31] Barnard, E.E. (1892) "Photographs of the Milky Way." *Astrophys. J*., 1:10.

[32] Hardie, op. cit.

CHAPTER 3

"Misty Wreaths and Streams of Filmy Light"

Barnard Goes to Lick

In 1887 Edward Barnard, now very widely known for his comet discoveries, accepted a position at Lick Observatory on Mt. Hamilton in California as one of the first astronomers hired at that new observatory. The director was Edward S. Holden, who came to be variously described by some of those who worked for him as "the Devil," "an unmitigated blackguard," "the Dictator," and several other unsavory epithets.[1] The social activities on the mountaintop, which grew to house some thirty to forty people, focussed around the astronomers and their families and the atmosphere developed into a cauldron of personality conflict and high psychological adventure. The center of the maelstrom was a conflict that grew between Barnard and Holden who would never see eye-to-eye on anything.

According to Donald Osterbrock, who has written a fascinating biography of Holden[2], the director realized that to staff his isolated mountaintop observatory he required a very special breed of person. The job description for the janitor, for example, was outlined by Holden in a letter; "He should have the manners of a lord, so as to please the visitors; he should be a good boxer, to keep order among the toughs; he should know considerable astronomy, so as to take care of the instruments, ditto chemistry—for the batteries—ditto arithmetic—to take meteorological observations. He must be a good housekeeper, always in good temper, understand how to chop kindling and to take care of our library etc., etc., etc.—and all this for $60 per month."[3] The advertising slogan for the U. S. Army, "Be all that you can be," describes the sort of person Holden was looking for!

Oblivious of what lay in store for him, Barnard and his wife travelled to California in early 1888 and arrived to "find matters in great confusion on Mt. Hamilton and no opening immediately available. He labored for several dark months in a lawyer's office in San Francisco copying legal papers."[4] Apparently this assignment boggled the imagination of friends who had tried to read his handwriting. One of then was later moved to note that he could "well imagine that the days were as dark for the lawyers as for the amanuensis. . . ."

During his seven years at Lick, Barnard lived on Mt Hamilton. It took six to seven hours to get there by stage from San Jose and five hours to get down again—not a viable commute. At first the couple lived in a shack originally used to shelter the workmen and in 1894 a brick house was completed. The Barnards moved in and quickly established the social hub of the mountain. They frequently organized picnics and on cloudy evenings had parties at their home. On cloudless nights Ed inevitably excused himself to work on one of the telescopes.

Barnard's life at Lick was described by a friend and colleague:

Barnard's most important work was the beginning he made during [the] summer [of 1889] in photographing the Milky Way with the Willard lens, which became a famous instrument in his hands. This was a portrait lens of 31 inches focal length, which had been used by some photographer and had received its name from the dealer who sold them in New York. This camera was strapped to a 6½ inch equatorial, which served as a guiding telescope. Barnard's long exposures with this instrument brought out the wonderful richness of the star clouds and other features of the Milky Way as they had never before been revealed. They thrilled him and his associates with their significance and beauty, and later the entire scientific world shared in this appreciation of them.[5]

From 1892 to 1985 he obtained many of his best photographs of the Milky Way with the Willard telescope (figure 3.1), but due to the difficulty of finding a suitable printing process, they were not published until 1913. Even then he only succeeded in publishing with funds obtained from friends of the Lick Observatory, and with his own money.

Figure 3.2 is a beautiful example of one the photographs taken with the Willard lens. It shows the dark markings near the star ρ Ophiuchi, a region of the sky that has become one of the most interesting to astronomers of modern times (see chapter 21).

Barnard's Experience Comes in Handy

His duties on the mountain began in June 1888, after which "he observed everything that shone and even things that obscured."[6] He became renowned for his acuteness of vision, but for five years was not allowed to use the large telescope thanks to the never-ending conflict with director Holden.

When Barnard went to Lick the timing was perfect. Photographic techniques had just evolved to the point where, with the advent of the dry plate process, it was practical to make astronomical photographs. Up to that time astronomical photography had been an exercise more of curiosity than practicality, but now Lick Observatory recognized the impor-

FIGURE 3.1. Barnard at the Willard telescope at Lick Observatory with which he made his fine collection of images of the Milky Way. He is wearing his "Esquimaux coat of reindeer skin." c. 1892. (Mary Lea Shane Archives of Lick Observatory)

tance of this new tool for research. Photography was about to reveal what no eye had ever seen.

It is within the experience of everyone who has used a camera that time exposures allow images to be recorded in faint light. The longer the

FIGURE 3.2. Region of the great nebula of ρ Ophiuchi. This modern print of Barnard's original 1895 plate is exposed to highlight the triple star that lies inside the ρ Ophiuchi nebula seen close to the center of the image. The entire region is filled with dark matter. The rings around the stars τ Scorpii (lower left), σ Scorpii (upper right), and Antares are due to halation, a phenomenon introduced in the photographic process. A broader view of this region is shown in Figure 4.4. (Lick Observatory)

shutter is opened the more light strikes the film and the brighter the image becomes. This is because the photosensitive material responds to the total amount of light reaching the film. This explains why astronomers were astounded when the first photographs of astronomical objects began to be produced. They saw things on the plates which had never before been seen; for example, details of nebulae, the dark markings, and comets previously unimagined. The plates also showed stars in enormous profusion where the eye had seen but thousands and revealed that the Milky Way contained vastly more stars than had ever been estimated.

For several reasons related to his boyhood employment at the photo studio, Barnard was ideally placed to pioneer the use of photography at Lick observatory. Despite his conflict with the director, he proceeded to do so.

Telescopes use mechanical tracking devices to follow a star or nebula, but the human factor was important to keeping the telescope exactly aligned on the guide star, usually tracked by watching through a small finder telescope attached to the large one. The better the telescope follows the motion of the heavens during the night, the clearer the photographic image obtained during the time exposure. If the telescope strays from its track the image becomes blurred. Who better in the world to keep a telescope moving accurately during long time exposures than Ed, the patient child who had spent hundreds of hours driving the cumbersome "Jupiter" to follow the sun. His sensitivity for this task, his capacity for appropriate feedback, was well honed.

He had another advantage. Unlike the vast majority of astronomers, Barnard knew a great deal about photography, printing, and enlarging. His experience with Poole's View Wagon now proved of inestimable worth. He was poised to launch a remarkable career in astronomical photography. And it was not to be free of great tribulation, thanks to "the Devil's" presence.

Psychological and Physical Hardships

By the time he reached Mt. Hamilton, Barnard's skills as an observer were unquestioned. However, comet discoveries, which had brought so much public acclaim, have never been regarded as very important by professional astronomers. As Osterbrock has pointed out, it was of greater importance to other astronomers of the age to compute an accurate orbit than to find another comet. Furthermore, since Barnard was not adept at arithmetic and could not calculate an orbit, he was not regarded as one of them, at least not at first.

Barnard recognized that despite his media success his reputation had not yet risen in the esteem of the professionals, and he became obsessed with achieving such recognition. He was intensely jealous in priority disputes which he judged as crucial to establishing his reputation, a syndrome that also haunts modern astronomers. Priority is often regarded as a basic ingredient in the hunt for career success, tenure, and grants. For Barnard priority became all important as a way to gain stature in view of the deep feelings of insecurity that followed him throughout life. "His childhood on the ragged edge of survival had scarred his soul deeply, and even at the heights of his career he could never really believe that some sudden stroke of fate might not leave him penniless and unable to continue research for which he lived. In his fifties, a widely-known professor, loaded with medals and honors, Barnard still worried that some day he might not have five dollars for his dues in the National Academy of Sciences, and that he might then be read out of the most prestigious scientific society in the nation."[7] Barnard was, in modern parlance, a highly neurotic individual.[8]

"When Barnard first came to Lick he was looking for a father Figure whom he could respect and emulate."[9] He was also, no doubt, seeking a guide and teacher, but was disappointed in the figure of the director on the mountain, who, he realized, "was not a skilled observer and was wasting the two nights per week that he assigned himself on the large Lick telescope."

Barnard's frustration rose as the "bumbling, ineffective former Army officer, posing as a scientist"[10] observed till around midnight and then retired to bed, leaving the telescope unused. At the same time Barnard was prevented from making use of it.

Why would Holden have labored so briefly at the world's largest telescope? A clue is garnered from his description of life on the mountain.

Before the least scientific work can be done, life must somehow be organized. If the shutters of the great dome are frozen together, the great telescope cannot be used. If there is no wood to burn in the office stoves, no computations can be made, no matter how enthusiastic the computer.[11] If the chimneys of the observatory will not draw, it is beyond any man's power to work at his desk, be he ever so devoted. The energy that is left over is available for astronomical work.[12]

We may well understand that after his daily trials as director of this mountaintop band of individuals he could not have had much energy left.

Holden also described the winter rigors on Mt. Hamilton and one wonders whether the astronomer's time would not have been better spent on exploratory expeditions to the North Pole. Figure 3.3 gives a sense of the conditions that caused him to write:

Out of six offices there are only two in a which a fire can be lighted in all winds. In one of the brick dwellings a fire will not burn in a southeast wind, and in the other in a north wind is equally fatal. The wind sweeps up the deep cañons on either side, and blows vertically down the flues, so that flames are driven into the room several *feet*! or else volumes of smoke make it simply impossible to

FIGURE 3.3. The astronomer's cottage at Lick Observatory on Mt. Hamilton, February 20, 1890. The optical astronomer's life is not all sunshine and roses! (Mary Lea Shane Archives of Lick Observatory)

remain in the apartments. Our meals have been served in the halls, in the bedrooms, or not at all![13]

A Hint of Something Lurking Between the Stars

In 1889 Barnard, who was about to become the unknowing pioneer of the study of interstellar matter, described his first photograph of the Pleiades (figure 3.4) with the following words:

The entire group of stars is filled with an entangling system of nebulous matter which seems to bind together the different stars with misty wreaths and streams of filmy light all of which is beyond the keenest vision and the most powerful telescopes.

This description was in terms of content and context, pure, classical astronomy. He did not tell us of the thrill he felt upon examining his photograph for the first time, which revealed much more than the eye, even aided by the telescope, had ever seen. (Scientists are remarkably reluctant to discuss the personal thrill that their work occasionally engenders.) In order for Barnard to understand what he had discovered he would require more than photographs. His "streams of filmy light" were indeed beyond the keenest vision, even with one's eye glued to a large telescope, and therein lies a great secret.

Photographic techniques bring out details invisible to the eye because the camera can make a time-exposure which allows faint light signals to be recorded. This the human eye can never do. Holding our eyes open

FIGURE 3.4. The Pleiades cluster of stars as photographed by Edwin Barnard in 1883 using a 3h 35m exposure. (Mary Lea Shane Archives of Lick Observatory)

for a long period of time does not cause the scene to appear brighter, or allow us to see fainter objects. This is fortunate, otherwise we would have a terrible time walking about with our eyes open. If they operated like a camera and film, the view would get brighter and brighter until we had to shut our eyes again to erase the accumulated image before taking another look. Photographic film, however, was to go where no human eye had ever gone before, far beyond the limits of human vision. It was there that the astronomical universe really began to reveal her secret self. The use of cameras attached to telescopes, a technique Barnard was to pioneer, marked the beginning of a great era of astronomical discovery. The subsequent study of light with spectroscopes followed and led to understanding the nature of the phenomena discovered by the camera.

Dark Markings Are Revealed, Recognized, but Not Understood

It was in August 1889 that Barnard began to make photographs of the Milky Way with the portrait lens (figure 3.2), a telescope that could perform this task as no other in the world. The director underplayed what Barnard achieved by way of early success, much to the latter's fury. Holden reported on photography of the Milky Way at Lick by stating that "Mr. Barnard had made some experiments in this direction in 1889, with the promise of most satisfactory results." This was not the way Barnard saw it! He had not experimented! He had taken three magnificent photographs of the Milky Way! These were in fact the first of a series that revealed the dark markings between the stars in gorgeous detail.

Barnard took Holden's comment as a "slur . . . on the Milky Way pictures I have made here."[14] He wanted the world to know about these photographs and proceeded to report on this work, insisting that it "seems desirable to give a brief description of the photographs . . . and to call attention to their special and important points which might otherwise be overlooked by those not familiar with celestial photography, and thus their value underestimated for the purpose for which they were made."[15] It is not difficult to imagine who the great unwashed were who were unable to appreciate his epochal photographs, persons "not familiar with celestial photography."

The art of astronomical photography was just beginning to evolve and the Milky Way had never before been captured in image. Barnard's first photographs did show "the wonderful and complex structure of the Milky Way,"[16] and he was proud of it.

Barnard Sees the Vacancies Between the Stars

The dark regions in the photographs of the Milky Way that Barnard began to find in many of his photographs had to be the "holes in the heavens," or "vacancies in space," that Herschel had claimed they were. And this is what Barnard decided to say when he published the first images in *Knowledge*, a popular science magazine, in 1894. This marked a very important point along the path that was to lead to the recognition of the existence of interstellar matter. The new editor of this magazine was A. C. Ranyard, astronomer and secretary of the Royal Astronomical Society. Following Barnard's brief description of one of the photographs, Ranyard made an important editorial comment which Barnard essentially ignored until much later in life. He did graciously acknowledge this, but unfortunately only after Ranyard had passed away.

Barnard offered a description of the image of the Milky Way around the star θ Ophiuchi which included one of the dark markings. He wrote,

"It will be noticed that in many of these vacancies there are 'deeper depths' yet, which almost suggest that the appearance of the diffused nebulosity over the region is real nebulosity, and that these dark and black patches in it are thin places and actual holes."[17]

Because he had read Thomas Dick's book (chapter 2), Barnard was aware of Herschel's beliefs concerning "holes in the heavens," and who was he, Barnard, to suggest that the dark regions were other than what Herschel had said? But Ranyard could not let the suggestion of vacancies among the stars slip by without critical comment. "The dark vacant spaces referred to by Prof. Barnard," he wrote, "seem to me to be undoubtedly dark structures, or absorbing masses in space, which cut out light from a nebulous or stellar region beyond them."

Ranyard did more than editorialize. He considered the geometry involved and pointed out that it is easy to imagine a narrow stream of dark nebulosity cutting out light and seen projected on the sky, but to believe there were holes which had to point away from earth in order to so neatly provide vacant space through the full depth of stars in those specific directions seemed impossible.

The probabilities against such a radial arrangement with respect to the earth's place in space seems to my mind to conclusively prove that the narrow dark spaces are due to streams of absorbing matter, rather than to holes or thin regions in bright nebulosity.[18]

Ranyard's criticism and suggestion were correct, but ignored; hardly surprising, since he was not presenting the idea in the proper international forum. All Barnard said about it at the time was that the photograph had been "accompanied by an article by the editor on the singular features shown in the photograph."[19]

We sense that he was not yet too interested in what his epochal photographs were revealing and must wonder where the study of interstellar matter would have moved if Barnard had from the start accepted the intuitive correctness of Ranyard's point of view and taken it to heart.

Barnard's Dilemma

Barnard's subsequent struggle to understand what his photographs of the Milky Way showed tell us that he was tortured by a profound dilemma. Was Herschel right? Were the dark markings (e.g., figure 3.2) holes in space? Or was Ranyard correct? Did matter exist between the stars? If so, he was privy to a discovery which would change forever what astronomers knew about space. But who was he, Ed Barnard, the poor unschooled kid from Nashville, to make a choice?

We must ask, in view of what was to follow, whether he really would have preferred to take beautiful photographs and not worry about what they revealed. Perhaps he felt that the interpretation of these images should be left to more learned colleagues. We will never know what he thought about this subject, because he did not tell anyone. What we will discover is that as the years went by he was tortured by this dilemma.

During the following years in his trays of fixer drifted magnificent images of a new phenomenon, something no one had ever dreamt existed; interstellar matter. But would he realize it?

Life on the Mountaintop

The drama of life on the mountaintop is revealed in chilly photographs of observers' cottages and telescope domes partially buried under giant snowdrifts (figure 3.3), which contrasted dramatically with the beauties of the summer which caused Holden to wax poetic: "Nothing is more charming than to drive to and fro; nothing seems (and nothing is) so delightful as a life of devotion to one's chosen profession among such beautiful and grand surroundings, as one of a company of fellow workers."[20] This was written from within the heat of personality conflict, by a man who may have perceived himself to be the epitome of the gentleman astronomer, performing one of life's gracious functions; observation and contemplation of the universe on its grandest scale.

Yet Holden was not beyond reporting favorably on the tireless work of his difficult employee, particularly in regard to a solar corona photograph that graced a Lick Observatory Eclipse Report. Barnard personally made 1,500 copies of this print and chose the 1,000 best, which he mounted himself to be bound as covers to the report—at a cost of only a few dollars, Holden once assured an audience.

The director still used only part of his two nights a week on the telescope, because he had to rest before the next day at the office. One can imagine Barnard fuming as he had to live beside the telescope which stood unused after midnight.

Barnard began to make his unhappiness at this state of affairs widely known, especially in face-to-face confrontations with the director, which inexorably led to a more bitter relationship. News of the feud leaked into the California astronomical community and the mountain top became a war zone between the director, on one side, and most of the astronomers, who lived in close proximity to one another on the little mountaintop, on the other. The saga is dramatically documented by Osterbrock,[21] and we sense from the exchanges between the antagonists that the tensions on the mountain were extraordinary.

Barnard's great joy came from observing, from the thrill of discovery, and then establishing priority of discovery, which was his very human need. He also gave many talks on astronomy to all kinds of popular audiences. He had to pursue his love for astronomy in an atmosphere of increasing conflict with the director who refused to let the self-made, famous astronomer use the large telescope. Thus while Holden frittered away the observing time Barnard craved, the situation on the mountain grew even more tense.

Barnard observed every clear night, always on the smaller telescopes, and had perpetual difficulty sleeping during the day. This created even greater stress for him. He never took a break and there was no one to insist he take a vacation, a task which had befallen his considerate friends at Vanderbilt, especially Bishop McTyeire. This fatherly cleric had once written to Barnard as follows, "You will oblige me, and everybody, if you will consent to take a *vacation* of two or three weeks. Even longer, if you find yourself away, and doing well. You need rest; if not now— you need rest to get strength for the future. And, certainly, you deserve it; you are entitled to it. The Stars and Comets will keep on their way, and be found in the right places when you return. Forget them for awhile. Dont look up, except to say your prayers, for the next month. Rest, *rest*."[22]

Don't look up! How could Barnard not look up? He was addicted to the heavens.

When Barnard's close friend, S.W. Burnham, who also had two nights per week to use the 36-inch, left to take another job, Barnard demanded that he be given this time. In a letter to the director he described that what he, Barnard, had done with the little time he begged from his friends was "among the finest and most important work ever performed with the great telescope."[23] This comment was not meant to assuage anyone's ego and was promptly turned down. Barnard broadened the attack. He appealed to the board of regents and communicated details of other squabbles to the chairman of the Lick Observatory Committee of the regents. He argued that he deserved the time and would finally be made to justify the telescope's existence. He also hoped that the director would be relieved of his post.

When nothing seemed to happen Barnard wrote to a friend, "I am sick of everything and everybody up here. I am going away. I am too sick even to wag my pen. Fraud is dominant. Falsehood flappeth his wings and croweth in triumph. I have fought the good fight and lost it."[24]

But he triumphed, partially. The regents interceded and in July 1892 he began regular observations with the 36-inch (figure 3.5). Holden, however, stayed on.

Barnard's Moment At Galileo's Side

On a Friday night, 9 September, 1892, Barnard struck pay-dirt. A simple observation brought him and Lick Observatory worldwide recognition. He had been looking through the 36-inch at Jupiter and noticed a tiny speck of light near the planet which, after more observations, turned out

FIGURE 3.5. Barnard at the 36-inch refracting telescope at Lick Observatory. He made this self-portrait shortly after discovering Jupiter's fifth moon which he called Jupiter V, later to be named Amalthea. (Mary Lea Shane Archives of Lick Observatory)

to be the fifth moon of Jupiter. He had been searching for such an object, looking for it closer in than the four well-known Galilean moons. No one had predicted it, but Barnard was interested in searching in any case. His report made his continuing frustration with Holden obvious.

Since July of this year, I have had the use of the 36-inch refractor on one night a week. Previous to this I had no regular use of the instrument, and the observations made with it were of specified objects, the time being limited to the object.[25]

He decided to name the moon Jupiter V and in so doing avoided any controversy that might accompany a more personal name. This satellite of Jupiter was the last moon in the solar system to be discovered by eye, which strengthened his reputation for being an excellent observer with extraordinary eyesight because in the 300 years since Galileo no one else had seen this moon. (All subsequent discoveries of moons orbiting other planets have been made photographically or from spacecraft.)

The exciting discovery captured the public imagination and brought instant fame to Lick Observatory and their new star, E.E. Barnard. Yet all was not well.

As a result of his battles with Holden, Barnard was often ill. Before the discovery of Jupiter V he had seen no future for himself at Mt. Hamilton. "My ill health is produced by worry," he wrote, "and I don't know to improve it unless I go jump in the Bay or clear out for the East."[26] One cannot help but marvel at this interesting choice of alternatives.

A Momentary Change of Pace

The discovery of Jupiter V (which was later named Amalthea) gave Barnard power and his friends urged him to fight to oust the director. At this time Barnard was given the prestigious Lalande prize of the French Academy of Sciences and was gone for seven months during which time Holden felt as if he, himself, were on vacation. When Barnard returned the fighting resumed and continued to be described in a deluge of letters—for the record. Barnard's psychosomatic problems often drove him to the sickbed from which he only arose to observe. Osterbrock, who has read hundreds of letters the famous astronomer wrote, concludes that "He was completely self-centered, hungry for praise, threatened by anything less than adulation, prey to all kinds of illnesses whenever his career was subject to pressure, a compulsive observaholic . . . his letters were all handwritten, and his style of writing changes within a letter, from . . . a not very well learned attempt at well formed letters when he was expressing conventional banalities to a childish scrawl when he was pouring out his grievances."[27]

Lest the reader think that Barnard's attitude to observing was unique, it is sobering to reflect that compulsive observaholics are still found at many observatories in the world.

In 1894 an offer came from George Ellery Hale (who went on to build Mt. Wilson Observatory and provide the impetus for Palomar Observatory) had been made director of the University of Chicago's Yerkes Observatory, still under construction. Yerkes was to have a 40-inch refractor, slightly larger than the one at Lick, and Hale very much wanted Barnard on his staff. The world-famous Lick astronomer was tempted but hesitated for several reasons. He didn't want to give up observing with the Lick refractor, he didn't like the idea of freezing in Wisconsin, and he was working at having his Milky Way and comet photographs published. The latter was going to be an expensive process and if he left for Yerkes the University of California might not pay for it. Finally he accepted Hale's offer, but wished it kept quiet until he had negotiated with the regents in California about publication, a move which resulted in complex political maneuvering behind the scenes. By now Barnard was important property, and his California friends did not want him to leave.

Barnard finally raised $2,000 for printing the pictures from a few wealthy friends but could find no company that was capable of printing them to his satisfaction. He held up the work for a decade until in 1913 improvements in photographic techniques satisfied him and the images appeared in print.

He finally left Lick under "unfortunate conditions,"[28] as one of his obituaries so politely phrased it. Roughly interpreted this meant that he left at the height of his conflict with "the Devil." His successor and the other astronomers finally brought the situation with Holden to a crisis point and forced Holden's resignation.

Notes

[1] Osterbrock, D.E. (1984) "The Rise and Fall of Edward S. Holden: Part I." *J. Hist. Astr.*, 15: 81.

[2] Ibid. And Part II. *J. Hist. Astr.*, 15: 151.

[3] Ibid.15: 92.

[4] Fox, P. (1923) *Popular Astronomy*, 31:195.

[5] Frost, E.B. (1923) "Edward Emerson Barnard." *Astrophys. J.*, 58:1.

[6] Ibid.

[7] Osterbrock, op. cit. 15: 94.

[8] Osterbrock, op. cit. 15: 92, an opinion based on reading hundreds of Barnard's letters (private communication).

[9] Osterbrock, op. cit. 15: 94.

[10] Ibid. 15:95.

[11] A *computer* in those days was a person, usually a woman.

[12] Holden, E.S. (1890) *Publ. Astron. Soc. Pac.*, 2: 50.

[13] Ibid.

[14] Ibid.

[15] Barnard, E.E. (1890) *Publ. Astron. Soc. Pac.*, 2: 240.

[16] Ibid.

[17] Barnard, E.E. (1894) *Knowledge,* 17: 253.

[18] Ranyard, A.C. (1894) *Knowledge,* 17: 253.

[19] Barnard, E.E. (1890) "On the Photographs of the Milky Way made at Lick Observatory in 1889." *Publ. Astron. Soc. Pac.*, 2:240.

[20] Holden, op. cit.

[21] Osterbrock, op. cit.

[22] Ibid. 15:95.

[23] Ibid. 15: 112.

[24] Ibid. 15: 113.

[25] Barnard, E.E. (1892) "Discovery and Observations of the Fifth Satellite to Jupiter." *Astron. J.*, 12: 81.

[26] Osterbrock, op. cit.

[27] Osterbrock, D.E. Private communication. June 25, 1987.

[28] Aitken, R.G. (1928) "Barnard at Lick Observatory. *Journal of the Tennessee Academy of Science.* Vol III, No.1. Edward Emerson Barnard Memorial Number: 20.

CHAPTER 4

"On the Question of Absorbing Matter in Space"

Veilings of Dust

Barnard arrived at Yerkes Observatory in the late summer of 1895. For the second time in his life his move brought him to an observatory not yet ready for his arrival. This time, instead of filling a temporary job, the famous astronomer went on a lecture tour and just before he left he heard that he had been awarded the Gold Medal of the Royal Astronomical Society, one of the highest honors an astronomer can receive. Andrew Carnegie paid for Barnard to travel to London to receive it. On such occasions the RAS holds a formal dinner for the recipient, but due to bad weather Barnard's ship arrived a day late and he missed the celebration. With British aplomb another dinner was organized for the very next day and the astronomer from the United States was suitably honored.

Yerkes Observatory was formed as part of the University of Chicago and was located at Williams Bay, Wisconsin. It was to have a 40-inch telescope, making it the largest in the world, but because it was not yet ready Barnard began to use a "magic lantern" lens of 1½ inches in diameter with a four to five inch focal length, well-suited for his photography of the Milky Way. In a report on the use of this telescope he stressed that "the most valuable and important information may be obtained with the simplest means."[1] Good advice then as now. With this lens he proceeded to make more glorious images of the dark regions and this would mark the beginning of his deep struggle to understand their nature.

Stories about Barnard's dedication to his work were to become legendary. He observed when the temperature outside, and hence in the observatory dome, reached 25° below zero. This forced him to stop work lest the telescope be damaged! Visitors would sometimes ask how the astronomers kept warm, to which he would reply, "We don't."[2]

In 1897 he again approached the challenge of the dark markings:

For many years this part of the sky [near ρ Ophiuchi, see figure 3.3] troubled me every time I swept over it in my comet seeking; though there seemed to be scarcely any stars here, there yet appeared a dullness of the field as if the sky were covered with a thin veiling of dust that took away the rich blackness peculiar to many vacant regions of the heavens.[3]

Despite A. C. Ranyard's suggestion made a few years before, neither Barnard nor the rest of the astronomical community was ready to embrace the concept of a veiling of interstellar dust.

In 1899 he described "long dull vacancies, running east and west" near the star θ Ophiuchi, as is shown in figure 4.1.

These peculiar dark apertures strongly remind one of the appearance sometimes presented in the umbra of sunspots, where darker holes lie in the dark central spot, as if the cavity were partly veiled with some sort of medium that itself had apertures in it—or a hole within a hole.[4]

He skirted any further discussion of the concept of a veil of material.

In Scorpio, he reported, "there is a slight suspicion that certain outlying whirls of this nebulosity have become dark, and that they are the cause of the obliteration of the small stars near."[5]

Barnard was now clearly torn. Were these holes between the stars or was some form of matter obscuring the view?

Wolf Knocks at the Door, but I Not Heard

Meanwhile, in Germany Max Wolf was about to come to the conclusion that the dark nebulae were composed of matter. Born in Heidelberg in 1863, Wolf had become very active in the use of astronomical photography in the discovery of asteroids and the study of nebulae, both galactic and those which were subsequently found to be extragalactic. His photographic survey of the Milky Way showed the same dark regions observed by Barnard, who appears to have been unaware of Wolf's existence. This was around the turn of the century, and neither man referred to the other's work, perhaps not surprising considering that communication was very slow at best.

Concerning a dark region in Cygnus, Wolf observed that "The most striking feature with regard to this object is that the star-void halo encircling the nebula forms the end of a long channel, running eastward from the western nebulous clouds, and their lacunae to a length of more than two degrees."[6] He asked, "is there a dark mass following the path of the nebula, absorbing the light of the fainter stars?"[7]

Wolf went on to write a report that demonstrated the existence of discrete clouds of interstellar matter. He used an elegant and simple quantitative test which relied on counts of the number of stars in each of a large number of brightness intervals as recorded on his photographic plates. In the absence of dark markings the cumulative numbers of stars counted would increase as one went to fainter magnitudes (figure 4.2),

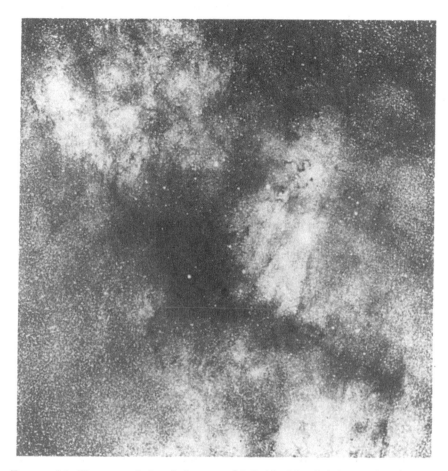

FIGURE 4.1. The great dark nebula east of θ Ophiuchi, a bright star lost in the bright area to the right of center. This print made from the plate taken by Barnard in 1906 reveals dark matter seen projected against the stars. Concerning these dark regions, Barnard wrote, "I think some of them will before long excite as much study and attention as the nebulae." He was correct. We now know that such dark clouds contain the very stuff of life. Notice the small, tilted S shape above θ Ophiuchi, labelled object 72 in Barnard's list of dark markings. The area reveals the inherent difficulty in defining dark markings as discrete "clouds." (Yerkes Observatory photograph, University of Chicago)

but in the direction of a dark marking he found discontinuous changes in the star counts, which indicated the presence of absorbing matter at discrete distances. These Wolf diagrams, as they came to be known, were the first proof of the phenomenon of interstellar absorption.

Barnard appears to have been unaware of this important quantitative approach to the problem of the dark markings. However, we must ques-

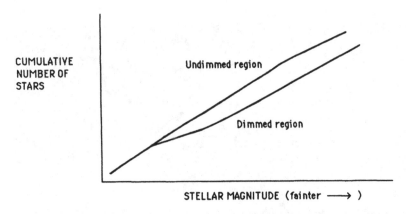

FIGURE 4.2. A schematic representation of a Wolf diagram. Star counts in a region suspected of containing obscuring matter are compared to a nearby, undimmed part of the sky. If no interstellar obscuration is present, the cumulative number of stars counted as one includes fainter and fainter stars increases at a steady rate. In the presence of obscuration (dust) at some distance the number of stars that will be counted drops off.

tion whether the technique would have proven anything when applied to some of the dark markings that concerned Barnard. A star count in their neighborhood would have revealed a sloping line such as shown in figure 4.2, but counts inside the markings would have given no data at all because no stars could be seen. This would not have settled the question of whether the markings were voids or indicative of the presence of matter.

With considerable foresight Barnard was to predict (in December of 1906) that the vacant regions, "will before long excite as much study and attention as the nebulae."[8] Yet this attention was still a long way in the future. In retrospect it appears that the study of dark nebulae was overshadowed by the explosion of effort in the more glamorous study of galaxies (formerly known as extragalactic nebulae).

The Bruce Telescope

In 1905 Barnard published the results of his photography of the heavens using the Bruce telescope at Yerkes Observatory. The telescope had been built in 1897 with $7,000 given to the University of Chicago by Miss Catherine W. Bruce, a benevolent, astronomically-minded spinster. Barnard, now Professor of Practical Astronomy at the University of Chicago, had personally solicited the money.

The Bruce funds provided for a photographic telescope of the highest quality, as Barnard's work was to reveal. He immediately set off on a trip to Europe (December 1899 to March 1900) in search of a suitable lens. It was a leisurely three month trip to let the strains of the work seep out, to lounge on the deck of the ship, and to travel by train on the Continent. We do not know whether he entertained fellow passengers with his knowledge of the starry heavens. Surely he would have. His trip is such a contrast to the movement of modern astronomers who must fly hurriedly across, carry out their business in a rush, streak back, and struggle with jet lag at both ends of the journey. Ah, for those gentler years!

In any event, his trip was for nought and he ended up using an excellent 10-inch lens made by Brashear, the same optician he had approached so timidly decades before. The mount was from Warner and Swasey and a 5-inch telescope was used for guiding. The device was completed in 1904 and first shipped to Mt. Wilson in California (in 1905) to view southern objects. An example of one of his Mt. Wilson photographs taken with the Bruce is shown in figure 4.3. This is the region around the star ρ Ophiuchi, which was previously photographed with the Willard lens at Lick (figure 3.2). The improvement in quality is dramatically obvious. The dark markings are more clearly visible in this wider angle view. When Barnard's observing session on Mt. Wilson was over the telescope was returned to Yerkes and housed in the Bruce dome.

Concerning the dark regions of the heavens, which were now even more clearly revealed, he was to write:

For many years I have called attention to the fact that many of the nebulae occupy vacant regions as if their existence was in some way the cause of the scarcity of stars. In reference to these dark lanes and holes, there seems to be a growing tendency to consider them dark masses nearer us than the Milky Way and the nebulae that intercept the light from these objects.[9]

He did not say among which group of people this tendency was growing, but briefly credited Ranyard for originally having put forward the idea of obscuring matter between the stars. Yet Barnard did not agree that such matter was involved. "I think they can be more readily explained on the assumption that they are real vacancies. In most cases the evidence points palpably in this direction."[10]

How excited Barnard must have felt when observing with the Bruce telescope. It allowed the heavens to be photographed with a clarity never before obtained. Had Catherine Bruce known about this work—she was already very old by the time she gave the money and may never have seen the results of her benevolence—she would surely have been delighted. Her donation was contributing materially to the discovery of new phenomena in the heavens. The habit of private benefactors supporting astronomical research is, unfortunately, all but nonexistent, with an occasional notable exception such as the benevolence of the Keck Foundation, which in 1984 made a substantial contribution for a major new telescope in Hawaii.

FIGURE 4.3. The dark lanes associated with the ρ Ophiuchi nebula, running east (to the left of the grouping of bright objects above the center of the image). "It is impossible to adequately describe in detail its extraordinary nature and that of the surrounding region," wrote Barnard. The object above the pentagon of bright objects, Barnard thought there was no finer nebula in the entire sky. The arc of bright stars and nebulous looking objects at the upper right, together with the bright stars below the nebula form the claw of Scorpio easily visible in the summer skies. The bright star at the lower left of the pentagon is Antares and the bottom right object is the cluster M4. (Yerkes Observatory)

One of Barnard's obituary writers later gave a delightful description of Barnard's work with the 40-inch telescope at Yerkes when he was not using the Bruce:

To him, a night at the great telescope was almost a rite—a sacred opportunity for a search for truth in celestial places. Rarely has a priest gone up into the temple with a deeper feeling of responsibility and of service than did this untiring astronomer go up into the great dome. He was usually ready before the sun had set, and patiently waiting until the darkness should be sufficient for him to "get the parallel" for the thread of the micrometer before he could observe fainter objects. During the day preceding one of his nights, his associates in the Observatory were generally conscious of his keen anxiety for a clear sky, as evidenced by a frequently repeated nervous cough, which was always worse if the prospects for the night were unfavorable.[11]

The Dilemma Confronted

Barnard's dilemma is most vividly evident in an article he wrote for the lay public in 1906. It was here that he stated quite firmly, "In most of these peculiar features the easiest and apparently correct solution is that these are real vacancies among the stars.[12]

He said he also had the impression that the most dramatic markings were seen where there "is a more or less thin sheet of stars with no great depth of thickness away from us."[13] This idea is startlingly incorrect. Stars are not found in sheets. They are spread throughout the depth and thickness of the Milky Way, but if they were in sheets he could understand why there might be vacancies or holes in such "sheeting," as he called it.

At this time Barnard also confessed his mental state vis-a-vis explaining why such vacancies existed. This problem was "another matter altogether, with which I have nothing to do."[14] This extraordinary confession makes us realize that he would rather have been observing and photographing and leaving the interpretation to others. But it was not to be. No one else seems to have cared very much, and in any case many may have been waiting for him to enlighten them. After all, it was on his photographs that the dark markings were so beautifully revealed. Did not he have the key to an explanation?

Barnard was obviously tantalized by the meaning of the dark markings but felt unable to delve into explanations for their existence. In view of what is now known about the markings, he was in any case ill-equipped to take on the challenge, as were most astronomers around the turn of the century. That was a time before the concept of the Milky Way as a galaxy had taken hold, before the nature of the luminous nebulae was understood, before the distance to stars beyond a few dozen parsecs could be measured, and before the existence of distant island galaxies was accepted. From this perspective Barnard's hesitation and inability to come to terms with what the dark markings indicated is more readily understood.

For the first time (in 1906) he also made reference to the fact that some of the dark regions are visible to the naked eye (the dark rift in Cygnus, for example) and groped at the idea that there were two types of dark marking. The first class were the very black "empty" spaces, seen within a dense "sheeting" of stars. This was intuitively reasonable, but only if he ignored why space beyond would not have more sheets of stars which would surely have caused the hole to be filled in again. The second class included "vacancies within vacancies" which he thought existed in Ophiuchus (figure 4.3):

... vast regions almost entirely free of stars, in a surrounding region thick with small stars. These regions seem veiled over with some sort of material in which occur blacker spaces, as if all this part of space were involved in a thin, faint, nebulous substratum which partly veils the blackness of space beyond.[15]

The rifts or openings were thought to give a clearer view of space beyond. Concerning the void east of θ Ophiuchi (figure 4.1) he was very specific: "No one would suspect for a moment that this lane is anything but an actual vacancy among the stars."[16] (He also recalled that these nebulosities were first found by him in 1883, while comet seeking in Nashville.)

Now Barnard was struggling to accept both concepts, true vacancies and veils of material. That it was a struggle is not in doubt. The extraordinary fact was that no other astronomer in the United States seems to have been seriously interested is joining him in his admittedly reluctant quest for an explanation. It is inconceivable in today's competitive world to imagine that a distinctly interesting phenomenon could be reported without a dozen astronomers jumping onto the bandwagon that would be quickly rolling in search of an explanation. (Success in such a search is widely regarded as a quick way to scientific immortality, perhaps deserving of a plaque in a corner of an imaginary Astronomical Hall of Fame.)

Barnard's dilemma was intractable because in order to solve the mystery of the dark markings he required more than photographs. He required quantitative data regarding the distances to the stars in the neighborhood and within the dark markings, and spectra to reveal the presence of absorption in the dark regions, as well as a host of other data which were only slowly forthcoming. Yet he persevered in his dilemma, frustrated by the lack of clear evidence for or against holes in the heavens.

... And It Worsens

The dark dust lanes in Taurus are famous examples of the phenomenon that so fascinated Barnard. "The vacant lanes are not only devoid of stars but ... they are clearly darker than the immediate sky. It will be seen that much importance depends upon whether these lanes are subjective—due to scarcity of stars alone—or whether they reveal to us a nebulous substratum in certain parts of the sky."[17]

This was the first clear hint that he considered that the lanes in Taurus at least were due to intervening matter, a substratum of some kind. Just as certainly as stars must die out in old age, so must the nebulae, and despite not knowing why nebulae shine, something that wasn't recognized for years to come, Barnard suggested that when a nebula does die it would become dark and invisible. It would then surely absorb starlight from beyond it.

Not only did he now recognize that dark matter might exist, but struggled to rationalize why. This was part of the ritual and preparation for changing his mind. He also realized that nebulae might not be transparent as comets were believed to be. This meant that when a star is seen in a nebula, one really had no idea whether that star's light was partially absorbed or not.

The idea of the absorption of the light of the stars by a dead nebula or other absorbing matter has been used by some astronomers as an explanation of the dark or starless regions of the sky. Though this has not in general appealed to me as a true explanation—an apparently simpler one being that there are no stars in these places—there is yet (considerable evidence) to commend it in some of the photographs."[18]

He obviously wasn't ready to change his mind and left the way open for exceptions, although his observations of the Taurus cloud had forced him, despite his hesitation, to consider the idea of intervening matter. In Taurus he could see parts of the darkness glowing faintly which hinted at the existence of a nearly invisible *nebula* as opposed to a vacancy between the stars. Bear in mind that the term nebula was used for the shining diffuse objects seen in various directions in the skies.

Returning to the dark markings in 1907, Barnard admitted that as far as the Taurus region was concerned, "I have been slow in accepting the idea of an obscuring body to account for these vacancies; yet this particular case almost forces the idea upon one as fact. There are portions of this apparent vacancy that are certainly darker than the adjacent sky."[19]

This was a crucial point. Barnard appears to imply that a simple vacancy would not be darker than surrounding sky. Why he believed this is hard for us to understand today. He now believed he had evidence for a substratum within which the lanes and holes are found and wondered whether the substratum itself consisted of nebulosity, which would be the natural conclusion. Or is it something else he asked, something "as to the nature of which we do not yet know anything." He was tempted by the notion of the substratum, but continued to shy away from it. "The idea of the dying out of the nebula is not strengthened by the presence of the lanes . . . ."[20] This notion seemed hardly possible from the appearance of the region when viewed through his telescope. He could not imagine a dead nebula as having the highly elongated shapes seen in the dark lanes in Taurus or Ophiuchus (figure 4.3). He thus retreated from the idea, stating that he did not believe in dead nebulae.

Kapteyn Makes His Mark

Around this time the famous Dutch astronomer in Groningen, J.C. Kapteyn, realized that the possible absorption of light would influence our understanding of the distance to stars. Born in a small village in the Netherlands in 1851, one of 15 children, Kapteyn grew to become one of the most well-known astronomers in the world and the observatory in Groningen is now named after him. During his lifetime he tirelessly gathered vast amounts of data in a variety of attempts to determine the structure of the universe.

Kapteyn suspected that there must be absorption due to matter between the stars, because space must surely contain meteoric matter, just like the stuff which falls to earth in the form of meteors. This matter would produce absorption and dim starlight which would cause us to believe the stars were further away than they really were. He had already observed fewer faint (i.e., distant) stars than he expected. This required an explanation which ran as follows: If the universe is completely transparent it appeared that the stars thinned out as we moved away from the Sun. However, if the star density were uniform this observation could only be explained by the presence of intervening matter absorbing starlight. Although he was correct in his conclusion, he was wrong in the premise. The star density in the heavens is not uniform.

Kapteyn's paper was nevertheless very important in raising issues which had to be considered. He was the first to coin phrases such as "absorption of light" and "selective absorption," the phenomenon that light of different colors are absorbed by different amounts as a result of the relationship between the size of the particles and the wavelength of light (of different colors). He also proposed that absorption due to gases might produce "space lines," that is, absorption over a very narrow wavelength band, to produce what are now called interstellar absorption lines, as discussed in chapter 17.

Absorption of starlight by interstellar calcium had been suspected since 1904 following the German astronomer J. Hartmann's discovery which is discussed later. However, Kapteyn did not refer to it in 1908, suggesting that even in Europe communication between astronomers was poor.

Concerning reasons why there should be gas between the stars Kapteyn stated flatly that

Owing to the gas of the corona lost by the Sun, to similar loss presumably suffered by other stars, to that lost by comets, etc., interstellar space must contain, at every moment, a considerable amount of gas. Might not this gas, in the thickness of hundreds of light-years, cause an appreciable absorption of light?[21]

It was a reasonable hypothesis but his question is now answered in the negative. Interstellar gas does not have its origin in stars in the sense suggested by Kapteyn, and the gas itself could not cause general obscuration of starlight. For that, solid particles are required. Gas would only cause spectral line absorption, as had been observed by then, but that observation was neither widely recognized nor understood.

Kapteyn pointed out that selective absorption had been reported by Antonia Maury in the USA, who had found that certain stars showed more absorption at the violet end of the spectrum. Annie J. Cannon, another U. S. astronomer, argued against the reliability of these data, but Kapteyn sided with Maury. He believed that to account for this "space absorption" one need only test whether those stars showing the effect were further away, a research project which was to occupy him for many years to come. But he already believed that the available data confirmed this idea. His longer-term goal was to get "insight into the true spectrum of the stars, freed from the changes brought about by the medium traversed by light on its way to the observer."[22]

Kapteyn suggested looking for space-lines or space bands which should be more pronounced in more distant stars. These lines would not share in the radial motion of the stars. Precisely this effect had already been observed by Hartmann and Kapteyn didn't mention it.

This famous Dutch astronomer's scientific papers stand out because of their broad sweep. Even if some of his suggestions have not stood the test of time, his were some of the first papers that ushered in what might be called a new era in astronomy. This was a true astrophysical analysis rather than a qualitative discussion of observations. In it Kapteyn concluded that there was no reason for assuming that the selective absorption is different for galactic and extragalactic stars.[23] (The notion of the existence of extragalactic stars was a sign of the times. It was the label given to stars located well away from the Milky Way, stars at high galactic latitudes in modern parlance. However, these stars do belong to the Galaxy.)

Kapteyn's results were discussed in a fascinating paper by George C. Comstock in 1910 which turned out to be a very odd red herring. This astronomer provided an ingenious but incorrect explanation of why the Milky Way is visible. He assumed the existence of two populations of stars, interpenetrating and permeated by "diffused meteoric matter, whose individual particles are of very small size and mass, but whose constant effect is to render less transparent the regions that they occupy."[24]

So far so good. But then he suggested that such particles should occasionally be swept up by stars. Thus a group of stars moving through space might sweep clean an entire region between the stars in another group creating maximum transparency and a greater richness of stars. From inside this latter group we would then see a band of bright stars all about us where space had been swept clean. In this manner Comstock accounted for the existence of the stellar Milky Way, a region devoid of dust. According to him the heavens well away from the Milky Way would be the most heavily obscured and hence darker. This is precisely the opposite of the true state of affairs!

By 1918 Kapteyn was confident that the amount of absorption of starlight was negligible and this forced him to consider that the sun was at or near the center of the Milky Way, an opinion which drew a lot of criticism. He persevered with his erroneous belief until he died in 1922.

In the meantime, though, other astronomers were discussing the existence of interstellar matter in a variety of forms, including calcium seen in the spectra of stars (see chapter 7). Barnard made no reference to these discussions or discoveries, but then he would have had little reason to associate gas between the stars with the phenomenon of obscuration of light which required a more substantial form of matter.

Barnard's Battle Revisited

In 1910 Barnard returned to the pages of the *Astrophysical Journal* with marvellous new photographs. The small lantern lens telescope had been brought to bear on a region of nebulosity near the star ν Scorpii. The title of his report included the words "on the question of absorbing matter in space." Now we hope that an irrevocable shift in opinion is beginning to take place. He saw fewer stars where a nebulosity was apparent and conceded that, ". . . the thinning out or dimming of the stars in this region, that are apparently in the nebula, is not due to a chance vacancy . . . the nebula is partially transparent but the absorption of the light of the stars behind it must be considerable."[25]

The concession had at last been made. In the same paper, referring to the nebulosity around ρ Ophiuchi (figure 4.3), he wrote, "the great nebula is located in a hole in a very dense part of the Milky Way, from which vacant lanes extend far to the east."[26] It is certainly easy to picture this hole when looking at Figure 4.3. The nebulosity can only be seen where we appear to be looking through a hole between the stars.

This paper was also historically interesting because of his unsuspecting use of a term now common in astronomy. He refers to a "small black hole in the sky, very much like a black planetary nebula,"[27] which he had previously observed in another part of the sky. "On account of its sharpness and smallness and its isolation this is perhaps the most remarkable of all the black holes with which I am acquainted. It lies in an ordinary part of the Milky Way and is not due to the presence or absence of stars, but seems really to be a marking on the sky itself."[28]

Now we grow more excited. Was Barnard at last ready to change his mind? Was he about to proclaim that Herschel was wrong and Ranyard correct? With bated breath we read on, "If these dark spaces of the sky are due to absorbing matter between the stars—and I must confess that their look tempts one to this belief—such matter must, in many cases, be perfectly opaque.... It is hard to believe in the existence of such matter on such a tremendous scale as is implied by the photographs."[29]

Come on now, E.E.B., we are rooting for you! How difficult it was for him to accept the enormity of what his photographs revealed. The existence of dark markings due to intervening material did imply tremendous amounts of interstellar matter, an idea whose time had not yet come. Once again he backed away. It was one thing to shock the astronomical community with a new discovery; another to add to that shock a conclusion which requires a substantial rethinking of a widely held belief. (Belief in the existence of a great deal of matter between the stars would raise other questions, and this implied all sorts of difficulties which were to concern later workers. For example, how would stars interact with this matter in their journeys through space?)

Barnard continued to be a purist, the unconvinced and super-cautious astronomer, wanting to be certain before he changed his mind. He referred to this darkness as being related to the nebulae which are known to shine. He asked, "Is it the ultimate condition of nebulous matter or is it something wholly different from the ordinary nebulosity of the sky?"[30]

Cautioning the reader who might be tempted to believe in the connection between the emission nebulae, such as seen in Orion (figure 1.3), and the vacancies, he stressed that he had also found vacancies in regions with no associated nebulae. "There exist dark lanes," he wrote, "which are so devoid of stars and nebulosity of any sort that an incident which befell Professor Tucker is worth recalling." Tucker had had the misfortune to point his telescope, unknowingly, at such an object. He had fixed his telescope and was watching the stars drift by. Barnard goes on:

One night he had set his telescope in the region a little north of Antares.... Presently no stars came into the field of his telescope. After watching for some time he finally concluded the sky had clouded over, but on looking out he found it perfectly clear. His telescope had been pointed to this lane and nothing but blank sky had passed.[31]

The dark lanes there happen to run east-west, so Professor Tucker's telescope had faithfully scanned along the empty lane (figure 4.3), a dramatic illustration of just how dark these lanes can be.

Despite the fact that in Europe other astronomers were beginning to suspect the existence of interstellar matter, Barnard's involvement in the study of the dark markings did not cease. Nor had his confrontation with his dilemma come to an end.

Notes

[1] Barnard, E.E. (1895) "Celestial photographs with a "Magic lantern" lens." *Astrophys. J.*, 2: 351.

[2] Calvert, M.R. (1928) "Some personal reminiscences." *Journal of the Tennessee Academy of Science*. Vol III, No. 1. Edward Emerson Barnard Memorial Number: 29.

[3] Barnard, E.E. (1897) "The great nebula of Rho Ophiuchi and the smallness of the stars forming the groundwork of the Milky Way." *Popular Astronomy* , 5: 227.

[4] Barnard, E.E. (1899) "Photographs of the Milky Way near the star Theta Ophiuchi." *Astrophys. J* ., 9: 157.

[5] Barnard E.E. (1906) "On The Vacant Regions of the Sky." Popular Astronomy, 14: 579.

[6] Wolf, M. (1904) "A Remarkable Nebula in Cygnus connected with Starless Regions." *Mon. Not. Roy. Astr. Soc.*, 64: 838.

[7] Ibid.

[8] Barnard, E.E. (1906) "On The Vacant Regions of the Sky." *Popular Astronomy,* 14: 579.

[9] Barnard, E.E. (1905) "The Bruce Photographic Telescope of the Yerkes Observatory." *Astrophys. J.*, 21: 35.

[10] Ibid.

[11] Frost, E.B. (1923) "Edward Emerson Barnard." *Astrophys. J.*, 58: 1.

[12] Barnard, E.E. (1906) "On The Vacant Regions of the Sky." *Popular Astronomy,* 14: 579.

[13] Ibid.

[14] Ibid.

[15] Ibid.

[16] Ibid.

[17] Barnard, E.E. (1907) "On the nebulous groundwork in the constellation Taurus." *Astrophys. J* ., 25: 218.

[18] Ibid.

[19] Ibid.

[20] Ibid.

[21] Kapteyn, J.C. (1909) "On the absorption of light in space." *Astrophys. J.*, 29: 46.

[22] Ibid.

[23] Kapteyn, J.C. (1909) "On the absorption of light in space—Second paper. *Astrophys. J.*, 30: 284.

[24] Comstock, G.C. (1910) "The absorption of starlight considered with relation to the galaxy." *Astrophys. J* ., 31: 276.

[25] Barnard, E.E. (1910) "On a great nebulous region and on the question of absorbing matter in space and the transparency of the nebulae." *Astrophys. J.,* 31: 8.

[26] Ibid.

[27] Ibid.

[28] Ibid.

[29] Ibid.

[30] Ibid.

[31] Ibid.

CHAPTER 5

"Dark Regions . . . Suggesting an Obscuration of Light"

Another View of an Idea Whose Time Had Not Come

In 1912 the U. S. astronomer Vesto M. Slipher made a remarkable discovery that was to be virtually overlooked for decades. If it had been recognized, the mystery of the dark markings might not have remained for long. Born on a farm in Indiana in 1875, Slipher was destined to make a name for himself through his measurements of the rotation and motion of galaxies in which he sowed the seeds for the expanding universe theory. However, it was his analysis of the light from the nebulosity near the stars in the Pleiades (figure 5.1) that should have excited the astronomical world. Instead, his profound insights fell on deaf ears and it is doubtful Barnard ever saw a report of Slipher's work, which would have illuminated his own problem.

Slipher found that the spectrum of the light from the nebulosity between the Pleiades stars was identical to that one of those stars, the bright Merope. Slipher had expected the nebulosity to produce its own spectral lines, which was true of other objects such as the Orion nebula (figure 1.4). But here the spectrum contained "no traces of the bright lines found in the spectra of gaseous nebulae." This puzzled Slipher, who noted that;

The light of this nebula which gives, technically speaking, a stellar spectrum, would naturally be accredited a stellar origin; but does the nebula shine by inherent light or reflected light borrowed from the neighboring stars?[1]

His point was extremely important. The nebulosity was clearly not stellar. To produce a stellar spectrum something was acting as a mirror reflecting starlight. Slipher discussed the possibility that the accumulated light from distant stars caused the glow of the Pleiades nebula but rejected the idea for various reasons, not least of which was a reference to the "well-known deficiency of faint stars in the Pleiades which has been attributed to the nebula's absorbing the light of the stars in the background."[2] He concluded that the nebula consisted of disintegrated (interstellar) matter similar to that known to exist in the solar system, and

that it shines by reflected starlight. This type of object became known as a *reflection nebula*.

With 20:20 hindsight we ask why no one made the connection between this phenomenon and the "vacancies," the dark markings Barnard had been photographing, which might be dark only because they contained no bright stars whose light might be reflected in the way it was in the Pleiades.

Slipher's work disappeared into the mists of obscurity which was a remarkable indication that astronomy was not yet ready to embrace the concept of interstellar matter. We will meet Slipher again much later, when his work was to be recognized after all.

A peculiar twist to Slipher's work was an extrapolation of his discovery of a reflection nebula to account for the existence of the Andromeda nebula (now known to be a large, nearby, spiral galaxy). Slipher believed that the Andromeda nebula, in a way similar to other spiral shaped nebulae, appeared to consist of "a central star enveloped and beclouded by fragmentary and disintegrated matter which shines by light supplied by the central sun."[3] He claimed this picture was consistent with the spectral data available to him, not unreasonable since the core of the Andromeda galaxy shines by the light of millions of stars. However, it was this starlight he was directly observing, not a reflection of that light.

Barnard Now Moves Ahead—Or Does He?

In 1913 Barnard again approached the issue of whether the dark markings were holes or obscuration and wrote about "Dark Regions of the Sky Suggesting an Obscuration of Light." The use of the cautionary word, *suggesting*, in the title warns us not to get too excited that he is about to make a breakthrough. His reminder that the work is only a suggestion is a widely used, conservative ploy adopted by scientists everywhere who are about to announce an important discovery, but one for which their colleagues may not yet be prepared! Barnard prefaced the report by noting that, "The so-called 'black holes' in the Milky Way are of very great interest."[4] This early use of the label, black hole, did not stick and should not be confused with the modern image.

The important point was that he at last conceded that obscuring matter was a more plausible picture than holes for explaining the dark markings. At last! we sigh. Barnard appears to have made the big leap.

◄────────────────────────────────

FIGURE 5.1. The Pleiades, as recently photographed at Lick Observatory. Wispy filaments of matter between the stars can be clearly seen. These are produced by starlight reflecting off interstellar dust. It is likely that Barnard's description of the "misty wreaths and streams of filmy light" in the Pleiades (Figure 3.4) was based more on what he saw through his telescope than what his photographic plate was capable of recording in 1883. (Lick Observatory)

We have witnessed his gradual conversion and felt his struggles as he considered a new point of view, one not even his to start with. What changed his mind? It turns out that on July 13th, 1913, he obtained a direct telescopic view of a dark object[5] and that,

... one would not question for a moment that a real object—darker looking, but very feebly brighter than the sky—occupies the place of the spot. It would appear, therefore, that the object may not be a vacancy among the stars, but a more or less opaque body.[6]

Heady words which gives us cause to cheer! But is our delight premature? Was he really converted?

His report showed one of the first photographs of the Horsehead nebula (figure 5.2) which, he stated, had not yet received the attention it deserved. (This may still be true.) This nebula is seen in projection against background nebulosity whose nature was unknown and it is difficult for us not to imagine that something dark exists in front of the bright regions.

The year 1913 also saw the publication of most of Barnard's collection of wonderful photographs, although the art of printing from the high quality plates was not yet at the level of technical perfection that would have kept this great photographer happy. He therefore added disclaimers to his description of each image to make sure the reader did not mistake some minor fogging with real astronomical phenomena. What was important, though, was Barnard's increasing tendency to confront the evidence for existence of interstellar matter revealed in this compilation of photographs.

Concerning the nebula associated with ρ Ophiuchi (figure 4.3) he now wrote

I think there is no question that strong indications of light obscuration are shown, though for various reasons the evidence is not so clear as in the case of the nebula about Nu Scorpii. One would hesitate in passing on the character of these dark lanes. The larger and more extended of these lanes is so blank and so definite on its edges that a long strip of obscuring matter lies between us and the stars here.[7]

"To me," he went on, in reference to a glow he observed in one dark notch in this region, "the observation would confirm the supposition of an obscuring medium." Could such statements leave any doubt? Such a glow was also in line with Slipher's discovery, but Barnard was unaware of that.

Others Begin to Speak Up

What seems extraordinary (in retrospect, of course) is that so few other astronomers exhibited any interest in the dark markings that Barnard was showing them; at least few if any seemed to join in the struggle to explain them. However, in 1914 another astronomer had commented on

FIGURE 5.2. The Horsehead nebula, one of the most famous dust clouds visible to astronomers. Barnard obtained one of the first photographs of this object, which is the thirty-third entry in his *Catalog of Dark Markings*. This image was obtained by David Malin with the 150-inch telescope. The wall of dust with its horse head outline is seen against a backdrop of matter illuminated by stars hidden from view. Barnard felt that this nebula had not received the research attention it deserved, which may still be true. (Copyright Anglo-Australian Telescope Board Telescope)

the nature of the dark spaces in the Milky Way. H. Knox Shaw wrote: "The dark spaces . . . are not those large regions conspicuous to the naked eye, such as the Coal Sack and the rift in Cygnus, but smaller regions which are found to be almost, if not entirely, void of stars, when viewed with a powerful telescope or studied with the aid of photography."[8]

Apart from this reference to dark regions visible to the naked eye, none of the astronomers of the time, Barnard included, ever suggested a connection between the dark lanes in Cygnus and the photographic dark markings. This seems extraordinary because the only difference is only one of scale. The two phenomena are manifestations of the same thing, clouds of interstellar matter cutting off light from distant stars. The Coal Sack (figure 1.1) in the southern heavens and the dark rifts in Cygnus are nearby (and thus apparently larger) examples of the objects photographed by Barnard. However, when one studies the dark lanes in Cygnus through a telescope stars immediately spring to view, thus the connection was perhaps not as obvious then as it is now. However, it is ironic that Knox Shaw so purposefully encouraged astronomers not to make the connection which would have helped them picture what was happening in the heavens.

This astronomer also reminded his readers that, "Sir John Herschel appears to have seen these and he thought the dark spaces were tunnel-like holes, the terminations of which were rendered nebulous by reason of their great distance."[9]

Knox Shaw was aware that Barnard was being forced to consider that at least some of these dark regions were due to obscuring matter and reported that Max Wolf in Germany was on the same trail. Wolf had noted that nebulae were often associated with the dark lanes and the remarkable nebula in Cygnus was apparently connected to starless regions which prompted Wolf to wonder whether it was "a dark mass following the path of the nebula, absorbing light of the fainter stars."[10]

He described this case in more detail:

A good instance occurs in Cygnus, which leads us to speculate whether it, or a cosmic process connected with its origin, has swept the long channel throughout the star clouds on the Milky Way. Or is there a dark mass following the path of the nebula absorbing the light of the fainter stars? He [Wolf] called this nebula the Cave nebula in Cygnus, and this cave is shown in his photographic plates to be full of even darker spots and channels.[11]

According to Knox Shaw, the Frenchman Courvoisier previously showed "that a vortex moving in a stream would leave such a path of dead water behind it as these lanes free of stars appear to be in the heavens."[12] It was difficult not to believe that some of the dark regions, a least those not associated with nebulosity, were really holes and thus he advised (as have so many astronomers since him in regard to a vast variety of problems) that more observations were required to resolve the issue. Regarding such observations, "there might well be many less profitable lines of work."[13] Parenthetically, we might add, there are also lines of work which have turned out to be demonstrably more profitable, especially outside astronomy!

Harold Spencer Jones, in England, who was to become Astronomer Royal, found independent evidence for the existence of obscuring matter revealed in photographs of edge-on spiral galaxies which often showed dark lanes running across the band of stars (figure 5.3). Sir Harold wrote, "A considerable amount of evidence is available pointing to a selective obscuration of light in space, and that the amount of this absorption—

FIGURE 5.3. An edge-on galaxy showing dark lanes of obscuration due to obscuring matter in its disk. An astronomer located inside such a galaxy would be unable to see through the dust and therefore would not observe galaxies in the plane of the galaxy (the equivalent of what we loosely label the Milky Way). This image of NGC 4565 was made with the 200-inch telescope. (Palomar Observatory Photograph)

although it cannot definitely be fixed—is undoubtedly very small, and probably does not begin to become important for distances less than 100 pc."[14] This was also in 1914.

Kapteyn again entered the picture in this year when he reported an observations that indicated that fainter stars are redder when they are further away, even though their spectral lines were the same. "Any distance effect must be explained by the presence of hydrogen in interstellar space . . .,"[15] he said, referring to the strengthening of the hydrogen lines in spectra of more distant stars. This had been observed by Walter S. Adams at Mt. Wilson and apparently reported to Kapteyn in a private letter. Today we know that it cannot be interstellar hydrogen gas that is responsible for the effect, but a result of the different spectral type of the more distant stars which are intrinsically more luminous. That is why they could be observed out to the greater distances in the first place.

The situation regarding the existence or nonexistence of matter between the stars became more confusing as a number of astronomers began reporting a variety of phenomena, all of which either suffered from obscuration or indicated the presence of interstellar matter and its effect on starlight. However, no one was able to put all the data together to form a coherent picture which would allow a breakthrough to be made.

Barnard in his next paper (in 1915) appeared to accept the presence of dark matter with very little comment.[16] Yet the next year he again left himself with an opening to revert to the old ideas. Despite our excitement at having seen him come around to the "correct" point of view, of winning the struggle with beliefs, he was apparently not ready to concede a change of heart. Once again he considered the existence of dead nebulae and said, "If we free ourselves from the belief that the function of a nebula is to become a star, or stars, we can imagine some similar condition to that of the star to prevail among these great bodies, that is, that they may become dark in the course of time."[17]

The Night Barnard First Saw the Light (or the Dark)

In 1916 Barnard repeated the argument that because stars are born and die, so luminous nebulae might have a similar life cycle, hence die out and become dark. He was now leaning heavily toward the idea that the dark markings were due to matter between the star, albeit dead nebulae. It was the general belief at the time that a nebula begins life in a luminous state and then develops into stars; "its ultimate destiny is a stellar condition."[18] Thus it was not unreasonable to ask what a nebula might look like once it ended its life. Still, no one came to his intellectual aid and he had to struggle on alone.

Barnard was now clearly drawn to the idea that there might be clouds of matter between the stars and in frank confession told the story of how the idea was brought home to him.

His epiphany had taken place more than a decade before when he was photographing the southern Milky Way at Mt. Wilson. It is astonishing that he took so long to confess his moment of insight, but as we shall see, the insight ran afoul of what he had learned from Herschel.

One night he was on Mt. Wilson, working with his beloved Bruce telescope. He had looked up at the bright glow of the Milky Way and seen something remarkable.

I was struck by the presence os a group of tiny cumulus clouds scattered over the rich star-clouds of Sagittarius. They were remarkable for their smallness and definite outlines—some being not larger than the moon. Against the bright background they appeared as conspicuous and black as drops of ink. They were in every way like the black spots shown in photographs of the Milky Way, some of which I was at the moment photographing.[19]

One can almost feel the idea that there were clouds of dark matter in space taking hold of him. The connection was made in his mind because terrestrial clouds had obscured the view. In those days the clouds would not have reflected city lights and so would have been totally dark. We wonder whether he leaped about, excited by what he was seeing and what his observation implied.

The phenomenon was impressive and full of suggestion. One could not resist the impression that many of the black spots in the Milky Way are due to a cause similar to that of the small black [cumulus] clouds mentioned above—that is, more or less opaque masses between us and the Milky Way.[20]

Imagine standing atop Mt. Wilson, looking at the dark heavens, the bright stars in the Milky Way, and then seeing a terrestrial cloud block the panorama in some small area. He mistook it for a dark marking between the stars and made the mental connection. The dark markings of his photographs might be due to interstellar clouds of dark matter!

A Seductive Association of Shapes

When Barnard made the connection in his own mind, what could he do to prove that it was significant? How could he prove his insight, his new model? He needed to make an attempt at a scientific argument ,which he proceeded to do, albeit it a decade after the insight first struck him on that clear night in the Californian mountains. His attempt failed.

Playing the cautious scientist to the hilt, he first admitted that there must be some obscuring masses in front of the stars. Then he drew attention to the objects shown in figure 5.4. One is a bright nebula in Cygnus known as NGC 6995, the other a dark marking in Cepheus. He could not resist making a connection. If NGC 6995 were to lose its light it would look exactly like the dark nebula. QED. It was a brave attempt to relate the two types of nebula, thus proving to Barnard that they were

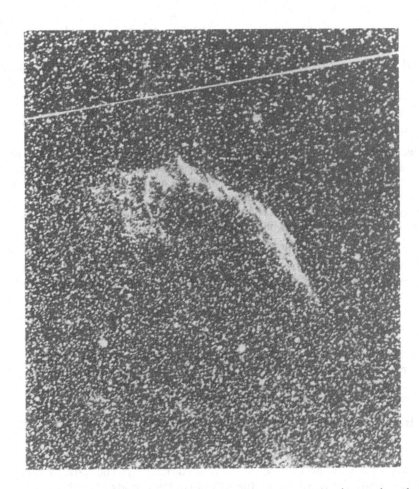

FIGURE 5.4. Comparison of a two markings on the sky, one luminous, the other dark. Their striking similarity in appearance led Barnard to thinking. If the luminous nebula lost its light, would it not look like the dark one? Thus he began to consider that perhaps the dark markings were due to obscuring matter between the stars, an idea he struggled with for decades. The left-hand image is of NGC 6995. The right-hand image has been artificially enhanced by Barnard, who made

both clouds of matter between the stars. But the argument did not work. He had compared only shapes. The cause of either the bright or the dark nebula was unknown, and comparison of two unknowns did not prove anything. That was only part of the problem.

In the early twentieth century no one understood why the luminous nebulae emitted light. Even less did they recognize that the bright nebula in figure 5.4 is the remains of an exploded star, a fragment of a supernova remnant. This was matter in the process of diffusing into space after being hurtled outward by the explosion. It is a form of matter utterly different

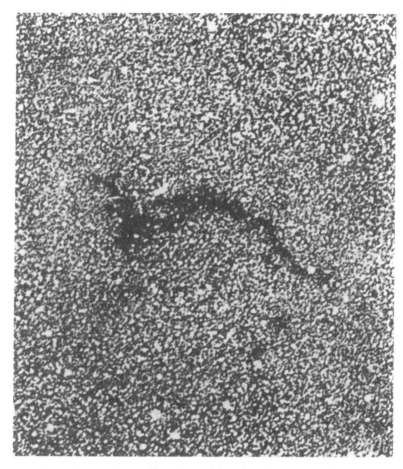

multiple printings with the position shifted a little each time, which gives the illusion of a dense star field. He did this to highlight the existence of this object, which was otherwise only very faintly visible, so faint that he suspected there had to be background emission against which it was seen (see text). The images were obtained with the Bruce telescope in 1910 using five to six hour exposures. (Yerkes Observatory, Photograph, University of Chicago)

from that seen in the Orion nebula (hot gas) or the Pleiades (cold material) or in the dark object shown in figure 5.4. But this would not be discovered for many decades to come.

The Sleepwalker

Barnard by now appears to be acting like the proverbial astronomical "sleepwalkers" described by Arthur Koestler in his book by that title. Astronomers of bygone years often confronted new truths about the nature of the universe without comprehending the relevance of what they had discovered and sometimes missing the discovery completely. Again, hindsight allows us to see this aspect of human nature in action and the point is made not to denigrate the great astronomers of the past, but to remind us to carefully and continually examine where it is we tread in our lifetimes. Thus watching Barnard at the edge of a major discovery, but unable to take the final step, becomes frustrating even as we hope he will ultimately make the final leap.

We will show later that the discovery of a new phenomenon, or the uncovering of a previously hidden truth about the universe, can be an extraordinarily thrilling experience. What can we say about those moments when one sees something new and doesn't recognize it as relevant? To make a discovery requires a prepared mind, and such preparation is based on education and experience, as well as bias, prejudice, and beliefs about what one expects to see. When Barnard saw the terrestrial cumulus clouds obscuring the light from the star clouds in Sagittarius he had an idea. "Perhaps clouds of matter actually exist in space, in front of distant stars," he must have thought. At last he could comprehend the way things hung together. But the ghost of Herschel was still with him, whispering into his ear-"holes in the heavens."

Barnard was aware that dark matter was seen in silhouette against certain nebulae which we know to be galaxies. For example, concerning the edge-on galaxy NGC 4565 (figure 5.3) he wrote,

Another beautiful example of (a nebulous strip seen in relief against a more luminous portion of the nebula) is shown in photographs of the very elongated nebula . . . NGC4565, which seems to be an object similar to the great nebula of Andromeda, with its edge toward us, where the darker outer periphery of the nebula is seen cutting across a brighter central region as a black irregular streak.[21]

How close he was to the correct explanation! NGC 4565 is similar to the Andromeda galaxy, but this is a flat spiral with its edge angled toward us. For him to have concluded this intuitive leap he would have had to become aware that NGC 4565 was a distant galaxy, but this was another idea whose time had not yet arrived. An edge-on spiral-type galaxy gives an exaggerated view of what we see from within our own galaxy, where the dark clouds lie close to the plane of the galaxy, the plane we call the Milky Way (see chapter 18).

Another Idea Is Tested

Barnard's struggle to change his mind continued. It was a struggle to which we can all relate when dealing with conflicting ideas and notions of a more local and personal nature, even if our concerns may be of a less cosmic nature than whether there or not vast quantities of dark matter exist between the stars.

Three years passed after he admitted that dark matter had to be present. Then, in 1919, he made a matter-of-fact statement to the effect that, "I think the general belief among astronomers is that a nebula remains luminous, and finally develops into a star or system of stars—that is, its ultimate destiny is a stellar condition."[22] He went on:

This opposes any supposition that a nebula may become dark by the loss of its light. For the presence of a dark nebula, however, it is not necessary that it should have lost its light. It may never have been luminous. It is possible that the original condition of a nebula is dark.[23]

Now we must sound our applause, for here was a marvellous insight very close to the modern view, however long it may have taken him to finally express it. This time he even suggested, very hesitantly, that the dark stuff might not be related to the luminous nebulae at all, "though this is perhaps far-fetched." He was again venturing into new territory and felt the uncertainty of a child wandering alone in the dark night of the unknown.

The Turning Point?

And so, after years of struggle with new concepts, it appeared that Barnard finally confronted the fact that the dark regions were not holes in the heavens. By 1916 he had settled on the recognition that obscuring masses do indeed exist between the stars. We feel relief and delight—for his sake. (From our safe perspective in the future, we wonder why it took him so long. We must ask ourselves how often we insist on clinging to our outmoded beliefs. Above all, what would we do if we knew which ones they were?)

In 1918 another astronomer, Heber D. Curtis, entered the debate by commenting on the photograph of the Horsehead Nebula (figure 5.2) as follows:

The most striking feature of the region is a remarkable dark bay jutting into and bifurcating the long ray. It is impossible to look at the original negatives of these interesting objects and not be convinced that there actually are "dark nebulae . . ." that they are "holes" torn in the star fabric of the Milky Way by some rapidly rushing star cluster is difficult to believe when one studies carefully the sharply defined edges. If merely "holes" we must assume their age as of the order of hundreds of millions of years, in which time, as Dr. Campbell has pointed out, the random motions of the stars in the neighborhood would long since have obliterated the clean cut edge, if not the "hole" itself.[24]

Barnard showed no indication of heeding Curtis's words and his new feelings on the matter become obvious in his 1919 magnum opus, a catalog of 182 "Dark Markings of the Sky". Figure 5.5 is a potent example from his plates. Eagerly we read to see if he will sum up. He does.

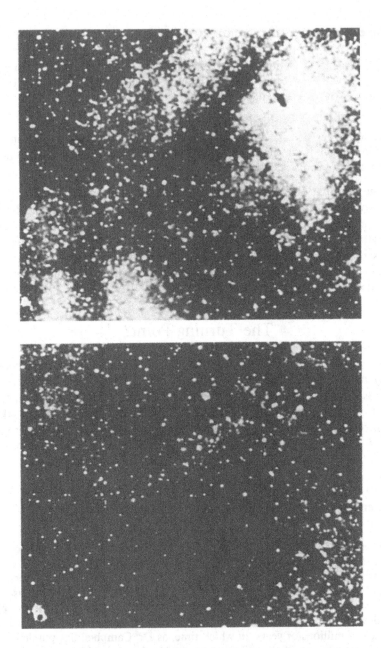

FIGURE 5.5. Lower image. Barnard #75. It is visible in Figure 4.1 just above the S shaped object. "It is a curious, narrow, looped black marking that covers about a degree in its peculiar windings." The upper dark marking is Barnard #84, which he thought was similar to #75 in having a scalloped appearance, not clearly visible here. However, these dust clouds are storehouses of interstellar molecules, undreamed of in Barnard's day. (Yerkes Observatory photograph, University of Chicago)

It would be unwise to assume that all dark places shown on photographs of the sky are due to intervening opaque masses between us and the stars. In a considerable number of cases no other explanation seems possible, but (that) some of them are doubtless vacancies.[25]

Oh no! we cry. Why? Why was it so difficult? After all this torture he still doesn't believe in matter between the stars! Was he experiencing sleepless hours in this struggle?

Why do people have such a terribly difficult time changing their minds? Perhaps it is always up to the next generation to change our minds for us, as Thomas Kuhn has pointed out.[26]

Unfortunately we will never know Barnard's innermost thoughts during his years of apparent transition, the years when he should have changed his mind for once and for all and said so. We do sense the frustration in his words which suggest that he suddenly forgot all his former doubts and concerns, "I do not think it necessary to urge the fact that there are obscuring masses of matter in space. This has been quite definitely proved by my former papers on the subject."[27]

Recantation

We now come to what may be the most dramatic part of Barnard's final "recantation."

I did not at first believe in these dark obscuring masses. The proof was not conclusive. The increase in evidence, however, from my own photographs convinced one later, especially after investigating some of them visually, that many of these markings were not simply due to an actual want of stars, but were really obscuring bodies nearer to us than the distant stars.[28]

By now Barnard had been looking at and photographing these objects for over a quarter of a century. Then he revealed again what had troubled him for so long. It was Ranyard's editorial in *Knowledge* in 1894, which first challenged Barnard's adherence to Herschel's notion of holes in the heavens, "Among the first to look upon these dark places as real matter was Mr. A.C. Ranyard whose lamentable death occurred December 14, 1894."[29]

Ranyard's important words had been:

The dark vacant areas or channels running north and south of the bright star [Theta Ophiuchi] at the center . . . seem to me to be undoubtedly dark structures, or obscuring matter in space, which cut out the light from the nebulous of stellar regions behind them.[30]

For twenty-five years Barnard had been troubled by his dilemma; was Herschel correct, or should he accept Ranyard's notion? Finally, when he had to change his mind, he gave the credit where he believed it was due.

And so we approach the end of his 1919 catalog of objects. Then, once again, we are stunned to read that he still believed that there are some truly starless fields in the heavens, voids in space. Why? This time because Sir John Herschel, William's son, had said so. Barnard had read in Webb's *Catalog of Celestial Objects*, based on the Cape observations of Sir John, that in the southern skies, according to Sir John, there appeared to be some cases of "real vacancies amongst the stars."

Notes

[1] Slipher, V.M. (1912) "On the spectrum of the nebula in the Pleiades." *Lowell Obs. Bull.*, Number 55: 26.

[2] Ibid.

[3] Ibid.

[4] Barnard, E.E. (1913) "Dark regions of the sky suggesting an obscuration of light." *Astrophys. J.*, 38: 496.

[5] At 1855 position, right ascension 18h 7m, declination $-18°$ 15'.

[6] Barnard, op. cit.

[7] Barnard, E.E. (1913) "Photographs of the Milky Way and Comets." *Publications of the Lick Observatory*, Vol. XI.

[8] Knox Shaw, H. (1914) "Dark spaces in the Milky Way." *Observatory*, 37: 98.

[9] Ibid.

[10] Wolf, M. (1904) "A remarkable nebula in Cygnus connected with starless regions." *Mon. Not. Roy. Astr. Soc.*, 64: 838.

[11] Knox Shaw, op. cit.

[12] Ibid.

[13] Ibid.

[14] Spencer-Jones, H .(1914) "The absorption of light in space." *Observatory*, 37:402.

[15] Kapteyn, J.C. (1914) "On the change in spectrum and color index with distance and absolute brightness. Present state of the question." *Astrophys. J.*, 40:187.

[16] Barnard, E.E. (1915) "A great nebulous region near Omicron Persei." *Astrophys. J.*, 41: 253.

[17] Barnard, E.E. (1916) "Some of the dark markings on the sky and what they suggest." *Astrophys. J.*, 43:1.

[18] Ibid.

[19] Ibid.

[20] Ibid.

[21] Ibid.

[22] Ibid.

[23] Ibid.

[24] Curtis, H.D. (1918) *Pub. Astron. Soc. Pac .*, 30: 65.

[25] Barnard, E.E. (1919) "On the markings of the sky with a catalog of 182 such objects." *Astrophys. J.*, 50: 1.

[26] Kuhn, T.S. (1970) *The Structure of Scientific Revolutions*, University of Chicago Press.

[27] Barnard (1919) op. cit.

[28] Ibid.

[29] Ibid.

[30] Ranyard, A.C. (1894) *Knowledge*, 17: 253.

CHAPTER 6

"A Self-Made Man"

A Lasting Legacy

"On the 6th of February 1923 he slipped quietly among the stars."[1] Thus wrote Philip Fox, an astronomer who the next day travelled by the 8.15 train from Chicago Northwestern Station to Williams Bay with several other professors to attend Barnard's funeral services. Frost described what he saw.[2]

He appeared pitifully wasted in body and was pitifully wasted indeed. With less furious pace he might have continued for some years his devoted service to the skies. I recall how last summer he said to me, "Oh! I'm so terribly lonely!" Who among observers has not felt this cruel loneliness, but I know he was thinking of his loss in death of Mrs. Barnard.[3]

His beloved wife had died in 1921, a loss which struck Barnard deeply and from which he never recovered. She had been his unfailing support in Nashville, during his years of tribulation on Mt. Hamilton, and at Yerkes. She had played the role of loving wife and gracious hostess to the full.

Frost recognized the inspiration that Barnard (figure 6.1) and S.W. Burnham, his friend and colleague, had been to others. "May we be worthy of these masters," he wrote in great sadness.

"As a child he witnessed the scenes of the Civil War which were enacted about his native city." Those were terribly times of poverty and under-nourishment for the mother and her children. Barnard even carried a scar from those times, visible on his right jaw from a sore that would not heal in his undernourished state.

"Mr. Barnard was stricken with diabetes early in the year 1914, and had to undergo the severe privation, by the doctor's orders, of giving up observations with the large telescope for a year. As a result of his obedience, his health was greatly improved, and for the past seven years [1916–1923] he kept up his observing most industriously and really beyond the measure of his bodily strength."[4]

Barnard received many awards, including the Gold Medal of the Royal Astronomical Society (1897), and the influential Sir Robert Ball called him the foremost observational astronomer in the world.

FIGURE 6.1. E.E. Barnard at the helm of the Bruce telescope. (Yerkes Observatory photograph, University of Chicago)

"He was an example of the possibilities which America offers its youth," wrote J. A. Parkhurst. "Nowhere else in the world could a boy rise from such restricted and difficult conditions to such complete and abundant manhood, honored equally for friendly character and high scientific attainments. The story of his early life is most romantic, though some of the features of his early youth during the war were so sad that he could not be persuaded to repeat them."[5]

Barnard once spoke of the great timidity with which he approached that great astronomer, Simon Newcomb, in 1877. Yet he rose to heights attained by Newcomb himself. We may all have a little of the Edward Barnard spirit within us. We have the potential to achieve what we strive for, even if it is a long time in coming.

It is remarkable and fascinating that the obituary writers, to a man, had no inkling of how important had been Barnard's studies of dark nebulae. Even in 1928 a popular astronomy text still reported that in regard to the dark regions "it is probable that a number of them do represent vacant spaces."

It has been said that Barnard was lucky in making the discoveries. These included the fifth moon of Jupiter, many comets, and hundreds of dark markings among the stars. In addition he obtained stunning photos of known comets and the solar corona during eclipses. The accusation of the role of luck could be made against any scientist who makes a breakthrough just because he happened to be in the right place at the right time. But as H. H. Turner said, luck may be involved, "but a man must not drop his catches."[6]

Concerning Barnard's buying of the 5-inch lens when he was struggling to support a family in Nashville, Frost wrote, "One can imagine what economies were necessary to accumulate the necessary funds and how firm the purpose behind the desire."[7]

When given the Gold Medal of the RAS, Dr Common said that "good sky, good telescope, good observer" combined to bring Barnard success. He was willing to work twelve or more hours of a winter night with temperatures at 20° below zero Fahrenheit, and "to sit patiently, even joyously, at his photographic telescope through an exposure of six or eight hours."[8]

One night at Mt. Wilson Barnard had made himself comfortable with his feet through the floor of the observing platform at the place where, unknown to him, a rattlesnake had chosen to make its den. This became apparent when the snake was caught in that very place. Barnard was not put off and suggested that the snake was there for the very purpose of warming the astronomer's feet.

After a night of observing he could be heard singing in the darkroom. "He was keenly appreciative of music, but not greatly gifted as a singer."[9] He also developed a love of poetry from his mother and was known to sing improvisations of "The Burial of Sir John Moore," a poem that had an astronomical error in it.

It was said that his only athletic exercise was golf, under protest, and an occasional swim in Lake Geneva, near Yerkes observatory. He was still a member of the Yerkes ski club (figure 6.2) when already well on in years. But it was widely acknowledged that the effort required to push the dome of the 40-inch about as often as he did was enough exercise for most mortals.

Barnard was a powerful swimmer, a skill that harked back to his days in the Cumberland River, salvaging food. On a swimming expedition to

FIGURE 6.2. Due to illness in his last years, Barnard was forced to cut back on his labors at the telescope. He was not averse to exercising and is seen here with the Yerkes Observatory ski club (far left). The photograph is believed to date from around 1920 when he was sixty-three years of age. (Yerkes Observatory photograph, University of Chicago)

Lake Geneva one of the professors had been warned that Barnard was not a good swimmer, and when Ed dived off the deep end of the pier "and came up blowing air through his mustache . . . ," this professor leapt in to attempt a gallant but needless rescue.

Burnham wrote. "Handicapped by the sorest distress and poverty from the first he has fought the battle of life alone and is in the purest sense of the word a self-made man."[10] He was genuinely loved by all who knew him, wrote Fox, "a most kindly spirit, most tenderhearted, making any distress of his friends his own distress . . .," and, "marked by great simplicity of character, great modesty, perfect unselfishness and self-abnegation." Homely views which contrast starkly with his battles with Holden on Mt. Hamilton.

Barnard had taken beautiful photographs which immeasurably enriched our knowledge of the contents of the Milky Way. To this day many of the dark clouds are known by a catalog number that bears his name. But he did a lot more than just photograph these mysterious markings. The work that made him famous was his observations of comets, double stars, Jupiter and its moons, and his estimates of the diameters of planets and their moons. He photographed the star clouds of the Milky Way and studied the bright nebulae. In all he won some eight major prizes that attested to his standing in the astronomical community.

Edward Emerson Barnard produced a lasting legacy in astronomy when he focussed awareness on the existence of the dark markings. However, the step of making a final unambiguous choice as to their nature was too great for him. It was left to others to make the break with the past and offer humankind a new concept, that of the existence of interstellar matter, a concept whose implications were to reverberate throughout all of astronomy. Sadly, Barnard did not live to appreciate that these dark nebulae carry the seeds of star formation and of life itself.

Notes

[1] Fox, P. (1923) *Popular Astronomy,* 31:195
[2] Ibid.
[3] Ibid.
[4] Frost, E.B. (1923) "Edwin Emerson Barnard." *Astrophys. J., 58:1.*
[5] Parkhurst, J.A. (1923) *Journal. Roy. Astron. Soc. Canada,* 17:97.
[6] Turner, H.H. (1963) *Astronomical Discovery,* Berkeley/Los Angeles: Univ. of California Press.
[7] Fox, op. cit.
[8] Ibid.
[9] Ibid.
[10] Ibid.

Part 2 The Evolution of an Idea: The Dilemma Resolved

Not long after Barnard's death, Arthur Eddington reviewed the evidence for the existence of matter between the stars and found the case to be overwhelmingly strong. A large number of astronomers began to take an interest in the study of both the solid form, interstellar grains, as well as the gaseous component.

Even if Edward Barnard had been convinced that dark matter existed between the stars he would have been hard-pressed to prove this unambiguously. The tools and techniques for doing so did not yet exist in his lifetime. Thus he lived on the knife-edge of a major discovery but was ill-equipped to make the conceptual leap that would have proved the validity of the idea that matter exists between the stars.

The scene is now set for the modern era in which the study of interstellar matter is to become one of the most active and exciting research areas in astronomy. The solid matter that drifts in ethereal clouds between the stars is the stuff of which planets are born. It is the stuff of life, the dust from which our earth and all upon it was built. It is to interstellar dust that most of our solar system will return in the inexorable course of astronomical time.

CHAPTER 7

"Astronomers . . . Must Presume the Existence of Such Matter"

Stationary Lines

Early in 1900 the French astronomer Henri Deslandres reported that he had discovered rapid changes in the position of spectral lines in the star θ Orionis. Astronomers at Potsdam Astrophysical Observatory in Germany were pioneers in building high-quality spectrographs, so this extraordinary discovery drew their attention. They had already discovered the effect of the earth's 30 km/s orbital motion on the spectra of distant stars. One of the astronomers there, Johannes Franz Hartmann, wrote as follows, "Director Vogel instructed the observers in the field of stellar spectroscopy at Potsdam to undertake to confirm the interesting phenomenon."[1] The picture of an observatory director *instructing* a team of observers to perform certain experiments is rare in modern astronomical circles.

Hartmann's 1904 studies disagreed with the claims of Deslandres. He found that the lines shifted only slowly and then found something even more extraordinary. The star Δ Orionis revealed the existence of "stationary" lines. Delta Orionis was a double star whose intrinsic spectral lines (the characteristic signatures of constituent elements in the star's atmosphere) showed periodic velocity shifts due to the star's motion about its companion. This was caused by the Doppler effect, which shifts the wavelength of the spectral lines by an amount that is proportional to the motion of the star. But Hartmann had noticed that in addition to the spectral lines due to the star, stationary lines which remained fixed in wavelength were also present. The gas producing the lines was calcium and its narrow H and K absorption lines, as they were called, had been observed (figure 7.1) by him. These showed a fixed Doppler shift independent of the motion of the double star and implied the presence of intervening clouds of gas somewhere between the star and the sun. The question was where; close to the star or in interstellar space?

Hartmann observed many other lines in the stellar spectrum and found that it was only the calcium lines at wavelength 3,934 angstroms[2] that

INTERSTELLAR LINES

Clouds of atoms in space make their presence known by their effect upon transmitted light. They absorb small amounts of energy from the starlight passing through them, thereby producing absorption lines in the spectra of the most distant stars. The strength of such interstellar lines depends upon the number of absorbing atoms lying along the line of sight, and their velocities within the atomic cloud.

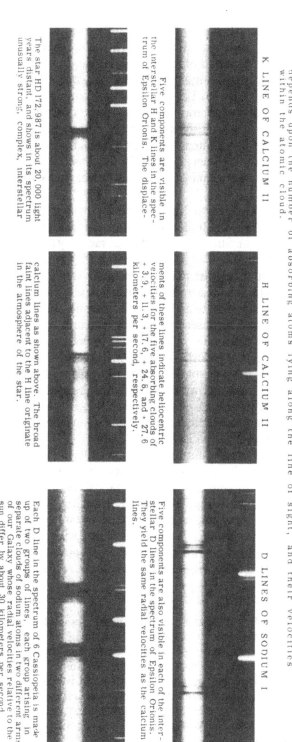

K LINE OF CALCIUM II

Five components are visible in the interstellar H and K lines in the spectrum of Epsilon Orionis. The displace-

H LINE OF CALCIUM II

ments of these lines indicate heliocentric velocities for the five absorbing clouds of + 3.9, + 11.3, + 17.6, + 24.8, and + 27.6 kilometers per second, respectively.

D LINES OF SODIUM I

Five components are also visible in each of the interstellar D lines in the spectrum of Epsilon Orionis. They yield the same radial velocities as the calcium lines.

The star HD 172,987 is about 20,000 light years distant, and shows in its spectrum unusually strong, complex, interstellar calcium lines as shown above. The broad faint lines adjacent to the H line originate in the atmosphere of the star.

Each D line in the spectrum of 6 Cassiopeia is made up of two groups of lines, each group arising in separate clouds of sodium atoms in two different arms of our Galaxy whose radial velocities relative to the sun differ by about 30 kilometers per second.

exhibited "a very peculiar behavior." He carefully ruled out several reasons for the stationary lines, including the earth's atmosphere, and concluded that an intervening cloud of calcium vapor had to be responsible. He noted that he had also seen this line in the spectrum of Nova Persei in 1901 and suggested that, "In the case of delta Orionis also it is not unlikely that the cloud (of calcium) stands in some relation to the extensive nebulous masses shown by Barnard to be present in the neighborhood."[3] This seminal comment was never seen by Barnard. Hartmann's discovery of the first direct evidence for the presence of interstellar matter of any form would remain essentially unrecognized for several decades.

In 1909 Edwin B. Frost detected the stationary lines in the spectra of several other stars and V. M . Slipher used more observations to support Hartmann's interpretation. Still, this news did not radically alter the course of astronomical history.

Slipher was the first to suggest that the stationary lines were actually due to interstellar matter, as opposed to material closely associated with, but not linked to, the stars. This had been Hartmann's implication. Slipher wrote, "May we not be observing in these calcium lines the 'space lines' which Kapteyn's researches have led him to predict, and the phenomena be due to selective absorption of light in space?"[4] Kapteyn was also on the trail of interstellar matter (as mentioned in chapter 4) and appears to have been unaware of Hartmann's results, although he was pleased to hear of Slipher's work. Kapteyn didn't regard the evidence as strong enough, however, to alter his own opinion about the nonexistence of widespread absorption due to interstellar matter.

Another of Slipher's ideas thus slid into obscurity, to be resurrected nearly two decades later. Thus is progress made, in a tortuous manner.

In 1920, R. K. Young found that the stationary lines only occurred in stars younger than B3.[5] This implied that the absorbing material had to be associated with the star, and hence had to be circumstellar. He concluded that the calcium was close to and associated with the star itself. It wasn't until 1926 that Sir Arthur Eddington[6] proposed a reasonable explanation for this effect. He suggested three reasons why the stationary

FIGURE 7.1. Examples of "stationary lines," absorption produced by interstellar calcium and sodium in the direction of a number of stars (see text). The light horizontal band is the optical spectrum in which wavelength changes along the band. Absorption at a certain wavelength due to specific atoms is manifested as a dark vertical "line." The bright vertical lines in the upper part of each diagram are produced by a reference light source at the spectrometer and used as calibration spectra, which allows the astronomer to identify which atoms are responsible for the observed lines. The production of absorption lines is discussed in chapter 16. (Palomar Observatory photograph)

lines were not seen in later-type stars. Firstly, for older stars the interstellar lines were masked by the broad spectral lines produced by the stars themselves. Secondly, the radial velocities of the other star types were generally not large enough to allow the stationary lines to be clearly separated. Thirdly, the older stars were fainter and in general closer to the sun, which suggested that they would have shown the effect. The more distant, older stars, were too faint to be observed and hence even if the "interstellar absorption" existed in their spectra, it could not be observed due to technological limitations. Eddington was to stimulate a lot of research on the topic (more about Sir Arthur will be presented later). Only then did the idea of the interstellar nature of the calcium clouds, as well as sodium, which had also been found to produce a "stationary" line, begin to gain acceptance. But not yet!

The possible existence of calcium clouds between the stars had excited F.J.M. Stratton who, in 1913, wrote:

If the existence of such clouds is ultimately well-established in different parts of the sky, it will be an interesting speculation for the stellar cosmogonists to derive from them all the information they may give. Will they, as possibly representing prestellar aggregations of gas, have no relative velocity and may they then be regarded as giving us a fixed origin and set of axes for our universe?[7]

This intriguing wish for an absolute set of axes by which to define such things as the "center" of the universe was in vain. Such an ideal would never be realized and in due course the calcium and sodium clouds soon revealed that there is anything but a fixed disposition of these objects in the galaxy. They showed motions with respect to the sun consistent with their being widely distributed in the galaxy.

Stratton wanted to ask more questions but realized the questions were in vain , at least until the clouds were a "well-established part of our known universe."[8]

By 1922 the possible nature of the material blocking starlight was alluded to by Henry Norris Russell who said; "it appears probable that the aggregate mass in one of these great obscuring clouds must be very considerable—probably sufficient to form hundreds of stars—at that a sensible fraction of the whole mass must be in the form of dust less than 0.1 mm in diameter."[9] This insightful statement went not much further because, as yet, he had no way to prove its accuracy.

Otto Struve Escapes and Makes His Name

The first report that really shouts of the existence of interstellar matter appeared in 1927[10] and was written by an extraordinary personality who had only recently arrived on the astronomical scene. This was the fourth generation representative of astronomers in his family, Otto Struve. His arrival was not without an extraordinary amount of luck.

Born in 1897, Otto was the grandson of the famous German astronomer F.G. Wilhelm Struve who had moved to Russia to superintend the building of Pulkovo Observatory. In 1919 the young Struve graduated from the University of Kharkov and in March of the next year found himself in Novorossisk on the eastern coast of the Black Sea, a lieutenant in the Imperial Russian Army. He was involved in the terrible war that followed the Russian revolution and was waiting to be evacuated and so escape certain death.[11] By the time he and his detachment of 300 reached the breakwater most of the rescue ships had already sailed. The weather was bitterly cold and many soldiers had died en route to the harbor. The captain of a Russian destroyer said he could allow half of Struve's detachment on board and then Struve had to stop the flow of men and together with the others remain in the port.

A sudden attack by the Bolsheviks caused the destroyer to fire back and in the confusion Struve managed to get all his men and himself on board and so they they were carried to Constantinople, Turkey. White Russian officers who had not escaped subsequently shot themselves before the Bolsheviks reached them.

Struve arrived in Constantinople with no passport, nowhere to go, and with the equivalent of fifty cents in his pocket. On a summer's day a year later he was to be swept up in a wondrous series of events when he met a Russian soldier he knew who happened to be carrying a letter for Otto. The letter was written in English which he could not read. So he bought a dictionary with half of his savings[12] but could still not translate the big words. He found help at "a building with a sign on it that looked as if it were written in English, even though he couldn't find the word YMCA in his dictionary."[13]

The letter had been written by E. B. Frost, director at Yerkes Observatory, who had become aware of Otto's plight through his uncle at the Berlin Observatory. Frost had a job for Struve if he could get to the United States.

The travel money was eventually raised by Frost with help from Princess Cantacuzene. She was the granddaughter of Ulysses S. Grant and had married a Russian prince. She knew who Otto was because she had been at court when Otto's grandfather was the Czar's astronomer at Pulkovo near St Petersburg.[14]

On October 7, 1921 Struve reached the United States and set out to become a famous astronomer in his own right. By 1923 he had obtained a Ph.D. in astronomy at the University of Chicago and much later was to become the first director of the National Radio Astronomy Observatory (in 1958).

Struve gained a reputation for being a fountain of ideas, some of which led to considerable controversies with his colleagues which he welcomed with enthusiasm, changing his ideas as rapidly as he learned more. The modern system for communicating astronomical results and ideas is too inflexible to allow such cavalier behavior!

Otto Sows Confusion, but Also Sees the Light (or the Dark)

In 1927 Otto Struve published his measurements on the intensity of the calcium absorption lines seen in the spectra of distant stars. His earlier work (in 1925 and 1926) had created considerable confusion because of his predilection to overinterpret poor data.[15] This first led him to suggest that the stationary lines actually moved and speculated that perhaps this was because revolving binary stars created eddies in the interstellar gas around them.[16] When this confusion, due to poor data, was recognized (the stationary lines do not shift their velocity), Struve claimed that the depth of the lines appeared to increase with increasing distance for the stars out to 500 to 600 pc from the Sun. It was there that the strongest lines were observed, he said. This result also did not stand the test of time.[17] Instead, the anomaly at 500 pc distance disappeared which caused Struve to suggest, "our final conclusion would ... be that the observational evidence does not contradict Eddington's theoretical ideas (for the existence of interstellar matter), but in some respects even favors them."[18] At this time (1928) Struve did notice the signature of galactic rotation effects in the velocity of the stationary lines, in accordance with the ideas which had, by then, been put forward by Oort (see chapter 17).

Struve's series of papers on the calcium lines did stimulate a tremendous amount of discussion on the possible existence of interstellar matter, although twenty years had passed since Hartmann first announced the discovery of these lines. In those intervening years, as Struve pointed out, knowledge concerning the origin of these lines had advanced little.

It is also remarkable that the notion of the possible existence of some cause of obscuration in space had been considered by his grandfather, F.G.W. Struve, who in the mid-nineteenth century had organized work at the Pulkovo Observatory on the distribution of stars.

He was able to substantiate earlier suggestions by Loys de Chésaux and Olbers that the interstellar medium is incompletely transparent and to determine the value of the obscuration effect in outer space, results that he published in the preface to Maximilian Weisse's star catalog of 1846 and in his own *Etudes d'astronomie stellaire* of 1847.[19]

This related to the problem of Olber's paradox. The issue of the presence of dark matter between the stars had been discussed for centuries as a way to (incorrectly) account for the fact that the sky was dark. The issue had been simple: If the universe were infinite, then wherever we look we should see stars and the sky would everywhere be aglow with starlight. Yet it is dark. Why? This was known as Olber's paradox. Astronomers and philosophers therefore wondered whether the universe was really only finite. Alternatively, the darkness might be caused by the obscuration of light from distant stars by intervening matter. This is not the cause for the dark skies. In its simplest resolution, Olber's paradox ceases to be an issue, because the universe is neither infinite nor do stars fill space evenly. They congregate in swarms called galaxies separated by the great starless voids known as intergalactic space.

If the elder Struve did manage to prove that there was obscuration of light in space, then that conclusion was forever lost upon the astronomical community, for it never surfaced again.

Otto Struve mentioned that Plaskett had suggested that "widely distributed tenuous clouds of matter" might envelop the stars so that the calcium would be ionized in the vicinity of the hottest stars, which, in turn, would be seen further away and thus give a wrong impression of the existence of this gas in interstellar space. But even this doubt did not last. In 1929, together with B.P. Gerasimovic, Struve wrote:

The hypothesis of an interstellar substratum embodying the whole galactic system is the most satisfactory at present. This hypothetical substratum shares the rotational motion of the stars around a distant central mass in (old) galactic longitude 325°.[20]

This direction in Sagittarius is now recognized as hiding the center of the Milky Way. They estimated the density of the material to be about 10^{-26} g/cm^{-3} (grams per cubic centimeter).

Sir Arthur Eddington Gets Involved

In 1926 Arthur Eddington entered the scene and proposed that interstellar matter should be widespread. When he gave the prestigious Bakerian Lecture to the Royal Society he chose as his topic, "Diffuse Matter in Space," the first technical review of the subject ever presented. The report makes for fascinating reading from several points of view: (1) for its insights, and (2) for its contrasting errors of intuition. This great scientist made several outrageous predictions which have never been born out. Many a student filled with trepidation about venturing out into the competitive world of real research might take heart from this fact and read the review.

The importance of Eddington's review is not in doubt. However, few have been willing to suggest that the paper had serious flaws, perhaps because Eddington has since been elevated to a stature larger than life. The review is made more fascinating because it reveals how a great mind like Eddington's makes progress through mistakes and an occasional chase up a blind alley which leads nowhere. Eddington was clearly fallible, and hence must be considered human!

Concerning the possible existence of interstellar matter (Figure 7.2 shows several examples), Eddington considered a point of view diametrically opposite (and hence more creative?) than had held sway for many centuries. He suggested that, "astronomers are . . . so placed that they must

FIGURE 7.2. A glorious example of the existence of interstellar dust, here seen in front of the Trifid nebula (M20) as photographed by the 150-in telescope in Australia. Hot stars have raised the temperature of interstellar matter to incandescence and caused the dust to shine in the lower region of the double nebula. Some of the light from the upper regions is reflected off dust, while the dark markings are due to molecule-laden dust clouds suspended in front of, and around, the nebula. (Copyright Anglo-Australian Telescope Board)

presume the existence of such matter unless its absence is proved"[21] (my italics). What an extraordinary statement to contrast with Barnard's struggles, Hartmann's misunderstood discovery, and Slipher's overlooked suggestions.

Eddington pointed out that up to that time the use of cepheid variable stars to determine distances to clusters and galaxies had been undertaken while assuming that no interstellar matter existed to dim the light. This was an important point, because the use of Cepheid variables required estimating the intrinsic magnitude of a star and comparing this with its observed magnitude. If unrecognized interstellar matter were present, then it would seriously affect the magnitude and hence any property that was derived from such data. Only when the effect of dust in dimming starlight was recognized could the Cepheids be used correctly to find distances.

Eddington included several other ideas in his talk which, in retrospect, were surprisingly eccentric. Among these was a suggestion that studies of stellar evolution implied that mass was lost by the star at the same rate as radiation. If this were so the star would have to accrete interstellar matter at the same rate in order to make up for the mass loss; otherwise the star would evaporate in a relatively short time. Thus, Eddington claimed, interstellar matter had to exist, if for no other reason than to feed matter back into the stars.

He also mistakenly (another judgement benefiting from hindsight) considered that the presence of an interstellar medium would control the dynamics (that is, movement) of stellar system more than did gravity. His point was that the resistance experienced by stars moving through this matter should slow them down. However, no such effect actually occurs because the density of the surrounding medium is far too low.

The thrust of these comments is to draw a perspective. Eddington, like so many theorists since his time, felt compelled to provide answers based judicious juggling of limited physical data and, when possible, of mathematical equations. Apparent consistency often follows from such jugglings because the human brain is infinitely clever in "explaining away" observed phenomena as well as unobserved phenomena. However, in many cases the consistency exists only within the mental formulation itself and not necessarily in reality. The exploration of possibilities such as Eddington indulged in is of course essential to progress in science and his examples reveal a great mind in action—mistakes and all. But it is this overdeveloped desire to explain things that gives rise to the cliché that in his description of the universe the theoretician is seldom hampered by data.

Eddington claimed that evidence for interstellar matter was to be found in the observations of dark nebulae, which often indicated a gradual thinning out toward their edges, suggesting that the matter existed in all of space and was merely concentrated in the dark structures. This marks the beginning of the concept of *cloudiness* in space, clouds of interstellar matter concentrated in regions with distinct boundaries.

When Eddington gave his talk in 1926 the existence of sodium and calcium in interstellar space was well established. Therefore he was able to estimate the density of interstellar matter, and this value was subsequently confirmed observationally by Struve. Eddington also suggested that the gas should be at a temperature of 10,000K, which is indeed the modern value for regions where the gas is ionized. But this is not the temperature of the typical region of interstellar matter (see chapter 20).

Eddington was the first to list a variety of ways in which energy is transferred from radiation to interstellar matter: through ionization of atoms, absorption during encounters with electrons, excitation of atoms (to produce spectral lines), and the scattering of free electrons. The first process, he suggested, would be negligible. It later turned out to be the most important. Bengt Strömgren was to point out that interstellar matter would be ionized by ultraviolet radiation from stars, with an ionized shell reaching to a distance which depends on the temperature of the star and the density of the surrounding gas. These shells of ionized matter around hot stars are known as *Strömgren spheres*. An emission nebula such as the one in Orion (figure 1.4) is really the combined Strömgren spheres of many hot stars in which the gas has been heated to 10^4 K.

While discussing the existence of sodium and calcium absorption lines, Eddington did not reference Hartmann's discovery, almost as if he wasn't aware of those data. In modern reviews the efforts of the pioneers in the field are traditionally mentioned, yet in this first ever review of the subject, little attention was paid to the most tantalizing observation of all. A decade later Eddington would rectify the oversight.

Flights of Fantasy

Eddington's Bakerian lecture is instructive and educational in many ways. It shows a pioneering scientist in full flight, but also illustrates just how far from the beaten path even the great sometimes stray. Eddington's first review is a good example of the dangers to be experienced when theory becomes master of the universe. The resultant flights of fancy may be interesting but not necessarily relevant. Sometimes a cleverly conceived theoretical structure can blind even the most ardent observer to the meaning of the data. This phenomenon is encountered time and time again in astronomy, in both ancient and modern times.

The Ptolemaic theory of the epicyclic motion of the planets and Sun ruled for nearly 2,000 years. It was elegant and wrong. For Copernicus or Galileo to question this, let alone propose a new theory, was no small challenge. Yet the Ptolemaic theory ruled the way astronomers functioned for nearly two millennia. Recently, upon suggesting to a young theorist that perhaps we would do well to bear this perspective in mind, that even the best formulated models might be wrong, I was met with, "No, not this time." Such idealism may sometimes overwhelm our perspective of reality. Some people in each successive generation believe that theirs is the one that has at last seen everything clearly, that their insights point to the truth, the final answer. Yet scientific discovery marches on and today's truths will become tomorrow's anecdotes.

The history of science is full of dramatic, and sometimes tragic, examples of this process in action. What can be more dramatic than Alfred Wegener's struggle to get geologists to consider the idea (theory) that continents drift? Wegener died before his ideas came to be recognized. Today the continental drift hypothesis is not only common knowledge, but measurements of continental drift have transformed the theory into a fact. Wegener, however, received not one iota of pleasure from realizing that others agreed that what he had suggested to be "true."

Eddington, in his epochal report on interstellar matter, next derived the amount of mass available in space, based on pronouncements made by a number of astronomers. He calculated the amount of scattering and absorption and related these to fully ionized atmospheres and then scaled the numbers to accord with conditions believed to exist in interstellar space. Theoreticians today would shudder at the extrapolation from stellar to interstellar densities, a range that covers a factor of about $10^{23.}$

To his everlasting credit Eddington realized that, "The only way of obtaining the obscuration with the mass at our disposal is by taking it to be in the form of fine solid particles."[22] This brilliant insight has since proved to be correct. Yet he couldn't accept it himself. He frankly admitted his bias against accepting the idea of fine particles in space. "I have great reluctance (which is perhaps a prejudice) to admit meteoric particles of this kind (a selected size range) in interstellar regions but I cannot suggest an alternative"[23] (his parentheses).

He further added to the printed version of his talk by admitting, "I was rather glad to find that my prejudice against meteoric matter was shared by others in the discussion on the lecture."[24]

Eddington's next statement turned out to be less relevant. He wrote, "The assumption that interstellar space is nearly transparent has usually been based on Shapley's observations that the light of the stars from the most distant globular clusters is not appreciably reddened. It is held that dimming without reddening is impossible (unless the obstruction is by solid meteoric particles of considerable size). In our opinion it is fallacious, since it applies only to un-ionized gas, whereas interstellar gas must necessarily be ionized."[25]

The last statement was presented with little doubt. Today we know that, on the contrary, it is neutral matter that is common in interstellar space.

The Voice of a New Authority

Some astronomers have favored Eddington as one of the pioneers of the idea of the existence interstellar dust, setting aside Barnard's mention of obscuring matter between the stars. Yet Barnard's suggestion was no more speculation than was Sir Arthur's. Eddington even said he didn't believe it. Yet his suggestion has been taken as more meaningful, perhaps because he made some rough mathematical estimates, since proven completely incorrect, regarding the amount of matter that had to be involved. This is a classic example of the tyranny of the theoretical speculation, disguised with numbers, however incorrect, that somehow adds mystique to any proposal. While a picture is worth a thousand words, a mathematical estimate is (unfortunately) attributed worth far greater than the eye can see.

Martin Harwit, author of *Cosmic Discovery,*[26] has recently explored the way certain astronomical concepts are hatched and then spread. He found that, "Concepts not actively championed by an authority figure are unlikely to gain recognition."[27] This now appears strikingly obvious as we examine how the notion of the existence of interstellar matter came to be accepted. Eddington's word was required, and a half-century of work done before he became involved was barely recognized at the time. Eddington's say-so at last made the subject respectable and the idea acceptable.

The Notion of Dust Is Introduced

Today the meteoric matter Eddington found so difficult to believe in is called *interstellar dust* and may or may not be related to the stuff that flies around inside the solar system, material left over from the era of planetary formation (see chapters 9 and 25). In Eddington's day the term *meteoric material* was all-encompassing. In order to avoid the idea that solid particles of any sort were involved, he noted that molecular absorption might produce the obscuration of light. He added:

It is difficult to admit the existence of molecules in interstellar space because when once a molecule becomes dissociated there seems no chance of the atoms joining up again.[28]

Eddington was far ahead of his time here, but still only partially correct. Interstellar molecules are common, and the reason for their existence is intimately associated with the "meteoric" material in whose existence he had so much difficulty believing. If molecules were to exist, he said, molecular absorption could account for the obscuration and he attempted to explain how this might work.

Possibly the temperature may be comparatively low in dark nebulae [it is] . . . In that case the atoms are mostly un-ionized [they are] and there is no obstacle

FIGURE 7.3. The η Carina nebula, one of the most beautiful objects in the sky, with dust clouds prominently obvious. The "keyhole" nebula is a dense dust globule seen in silhouette in front of the luminous gases. The direction of the strikingly beautiful constellation Carina in the southern skies is otherwise remarkably free of interstellar dust. It is possible to see stars more than 10 kiloparsecs distant in the Milky Way in this direction. (Copyright Anglo-Australian Telescope Board)

to the formation of molecules [modern theorists would argue that] and consequent molecular absorption. But aught we not in that case to find some traces of band spectra in stars which are partially dimmed by the nebula? [Molecular line absorption was later found.][29]

Eddington's talk was clearly a watershed. It suddenly raised the study of interstellar matter to a respectable level, because of the sheer weight of his reputation, and his discussion of the subject was to influence a great many young astronomers of the time.

The possible origin of the nebulae was delightfully painted by him, beginning with a uniform distribution of interstellar matter which is swept up by stars until "all the condensations are happily settled with no matter left in holes." Irregularities in the shapes of the nebulae are pictured as being the result of "superfluous matter bandied about from one (concentration) to another."

His was the first attempt to describe the formation of stars out of the nebulae (figure 7.3), a daring attempt which would be developed and redeveloped not only by him but by hundreds of other astronomers in the years to come. For the student of interstellar matter the original Eddington picture is intriguing, given the lack of high quality data available in those days. The picture he painted turned out to be very different from the modern one. This is not a criticism but a plea to young astronomers to recognize that history may look back on the work of our era and recognize that we, too, may have been naïvely—perhaps even quaintly—misguided in our beliefs. And so progress is made.

Notes

[1] Hartmann, J.F. (1904) "Investigations of the Spectrum and Orbit of Delta Orionis." *Astrophys. J.*, 19:268.

[2] An angstrom is the optical astronomer's unit of wavelength. $1 \text{ Å} = 10^{-8}$ cm.

[3] Hartmann, op. cit.

[4] Slipher, V.M. (1909) "Peculiar Star Spectra Suggestive of Selective Absorption of Light in Space." *Lowell Obs. Bull.*, 2:2.

[5] Stars are classified according to a scheme which ranks their intrinsic luminosity which is roughly related to the star's mass and age. The sequence O, B, A, F. G, K, M is one that proceeds from massive, luminous, young stars to smaller, less luminous, older stars.

[6] Eddington, A.S. (1926) "Diffuse Matter in Space." *Proc. Roy. Soc.*, A111:424.

[7] Stratton, F.J.M. (1913) "Some Problems in Astronomy. IX Fixed Calcium Lines." *Observatory* , 36:366.

[8] Ibid.

[9] Russell, H. N. (1922) *Proc. Nat. Acad. Sci.*, 8:115.

[10] Struve, O. (1927) "Interstellar Calcium." *Astrophys. J.*, 65:163.

[11] Sweitzer, J.S. (1987) "A Most Exceptional Star: The Life of Otto Struve." *Griffith Observer*, September

[12] Ibid.

[13] Ibid.

[14] Ibid.

[15] Berendzen, R., Hart, R., Seeley, D. (1976) *Man Discovers the Galaxies.* New York. Science History Publications.

[16] Struve, O. (1925) "On the Calcium Clouds." *Pop. Astr.*, 33:639 and (1926) *Pop. Astr.*, 34:1

[17] Struve, O. (1925) "Further Work on Interstellar Calcium." *Astrophys. J.*, 67:390.

[18] Ibid.

[19] From *Dictionary of Scientific Biography* p. 110 Charles Scribner and Sons, New York, 1980.

[20] Gerasimovic, B.P., Struve, O. (1929) "Physical Properties of a gaseous substratum in the galaxy." *Astrophys. J.*, 69:7.

[21] Eddington, op. cit.

[22] Ibid.

[23] Ibid.

[24] Ibid.

[25] Ibid.

[26] Harwit, M.O. (1984) *Cosmic Discovery,* Cambridge, Mass.: MIT Press.

[27] Harwit, M.O. Private communication. December 11, 1987.

[28] Ibid.

[29] Ibid.

CHAPTER 8

"I Wonder What Our Successors Will Have Made of This Great Problem"

An Idea Whose Time Had Finally Arrived

In 1930 the question of absorption of light in interstellar space was laid to rest by Robert Trumpler[1] and in the process the concept of the existence of interstellar matter finally came to be accepted throughout the world. Trumpler used a simple, elegant method to show that such absorption occurred because of the drop in brightness of distant stars whose spectra were otherwise not different from nearby stars. They were also redder than normal. The extinction of the light from distant stars was found to be concentrated in the Milky Way.

Forty years had passed since Edward Barnard had first photographed the "vacancies in space." A century had passed since William Herschel voted for the existence of interstellar holes in the heavens, and two centuries since the concept of gaps between the stars was first aired by Ferguson. Now the vacancies, holes, and gaps would all be filled with matter in a variety of conditions.

Born in Zurich, Switzerland, in 1886, one of ten children, Trumpler emigrated to the United States in 1913. He became the first astronomer to list the many ways that light could interact with matter on its way through space, a list which remains essentially unaltered, although somewhat extended today:

General Absorption of Starlight

Light at all wavelengths decreases more rapidly on its journey through space than the inverse square law predicts. This is due to the presence of an absorbing medium of (dust) particles which blocks the light and affects the interpretation of stellar distances based on measuring stellar light levels. In some directions, such as Sagittarius, the absorption is total and no light from stars more than a few thousand parsecs away reaches earth.

Selective Absorption or Reddening of Starlight

This is a relative effect manifested across the spectrum. Light at the shorter, blue wavelengths is absorbed more than longer-wavelength red light, an effect expected if the absorbing particles have a characteristic size comparable to the wavelength of light (5×10^{-5} cm). A similar reddening effect is seen on Earth at sunset, when the sun shines through a great depth of dust near the horizon. The shorter-wavelength blue light is scattered and colors the sky blue. The longer-wavelength red light travels through the haze before reaching the eye and thus colors the sun red.

Spectral Line (Monochromatic) Absorption

Light at discrete wavelengths is absorbed by atoms or molecules in interstellar space, with the spectral lines observed being the signature of the various species of the atoms in question. In Trumpler's day only atomic species such as calcium and sodium had been seen in space. He stated that the work of Struve, Plaskett, and Eddington demonstrated that this was a real effect. The "stationary" absorption lines were at last firmly attributed to matter in space, not in stellar atmospheres. Trumpler failed to mention the discoverer, Hartmann.

Obscuration Effects

The presence of interstellar matter could be seen by looking at photographs of the Milky Way, which revealed the presence of dark nebulae that cut off light from more distant stars. This was taken to be so obvious that no reference to any specific work was given, which paints Barnard's struggle in a sobering light.

Another dramatic demonstration of obscuration had been found in the way galaxies and globular clusters (figure 8.1) are spread over the sky. They avoid a band close to the galactic plane, a band which later became known as the *zone of avoidance*. As early as 1919 Henry Norris Russell believed that the existence of dark clouds in the Milky Way accounted for this effect and said so in a private letter to his student, Harlow Shapley, who had found no other explicit evidence for absorption in the direction of globular clusters.[2] He expected to find this absorption if Russell was correct.

Heber D. Curtis had (in 1917) invoked the existence of obscuring matter to relate the zone of avoidance to similar bands found in the photographs of edge-on spiral nebulae (fig. 5.3). Those galaxies show dark bands of dark matter in front of a stellar realm. Curtis also suspected that this obscuring matter prevented a view of globular clusters in the plane of the Milky Way galaxy because we were trying to see through so much material there.[3] Nature has distributed the obscuring matter in the disk of the galaxy and not in the directions of the majority of globular

FIGURE 8.1. The globular cluster 47 Tucanae photographed by David Malin with the 150-inch telescope in Australia. Hundreds of thousands of stars are locked in gravitational embrace in such clusters. (Copyright Anglo-Australian Telescope Board)

clusters which lie at high galactic latitudes (see chapter 17). The globular clusters, in turn, fill a great spherical volume of space around the galaxy. The two discrepant points of view are now resolved, but back in the early twentieth century they gave rise to a lot of bickering.

Shapley and Russell were to argue this confusing issue for years.[4] Shapley objected to Russell's views, and over a period of four years made observations of the locations of globular clusters[5] which seemed to support his objections. None of the globular clusters lay closer than 1 kpc to the galactic plane and this therefore would give the *illusion* of a zone of avoidance (see figure 8.2). He went so far as to suggest that the disruptive effects of the galaxy's gravitational pull prevented the formation of the globular clusters in the plane.[6] Russell wasn't convinced. Nor was the influential George Ellery Hale, who cautioned Shapley "on the dangers of making daring hypotheses without presenting proof."[7] Shapley sent Russell another diagram which included the location of Cepheid variables and open (galactic) clusters, and Russell noticed that these are not seen farther than about 17 kpc from the sun.

The two effects are now accounted for in terms of obscuration in the disk cutting off the view to more distant open clusters[8] and Cepheid variable stars. Also, the globular clusters are not found in the disks of galaxies, representing, as they do, a very different "population" of older stars.

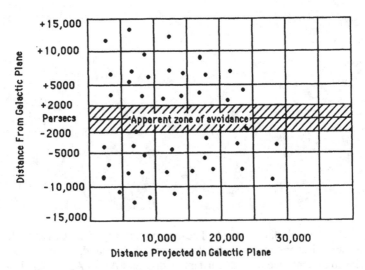

FIGURE 8.2. Schematic representation of Shapley's diagram, showing how the distribution of globular clusters with respect to the Milky Way could give the illusion of a zone of avoidance if the clusters were a population of objects swarming around the Galaxy. Shapley was able to plot the actual locations of the clusters and found them to avoid the disk of the Milky Way.

The argument raged with Russell (correctly) convinced that "one can almost see the absorbing matter in the galactic plane, cutting off the remoter open clusters from view."[9] Regarding properties of the interstellar matter, Russell also correctly guessed that "dust or fog is a far more powerful absorber per unit mass than gas, and most of the matter in interstellar space ought to be solid—if its chemical composition is in the least like that of the stars—but may be finely divided."[10]

In 1926 Edwin Hubble published his observations of the distribution of galaxies over the sky and the zone of avoidance was even more clearly delineated. It is impossible to see galaxies through the dust layers in the Milky Way, although radio waves pass through this material quite freely. In the last few years dozens of previously unknown and relatively nearby galaxies have been found in the zone of avoidance, which covers a quarter of the sky. They were found by Frank Kerr and Patricia Henning of the University of Maryland through the detection of the radio waves emitted by interstellar hydrogen gas within the galaxies.[11]

Further Evidence for Interstellar Matter

Before we examine Trumpler's proof of the existence of interstellar matter it is appropriate to complete the list of ways in which such material is manifested.

Polarization of Starlight (Discovered in 1948)

In passing through dust clouds, starlight becomes polarized—i.e., those light waves vibrating parallel to one particular plane are passed more freely than those in other directions—as a result of the interception and extinction of waves by elongated dust particles that have been aligned by the magnetic field in the interstellar cloud.

Emission and Reflection Nebulae (1880 to 1900)

The former are produced when young, hot stars heat interstellar gas to such an extent that it emits heat (infrared), light, and radio radiation; e.g., the Orion nebula in figure 1.5. The latter are dust clouds lit up around cooler stars and shining by reflected starlight, much as fog or mist glows around street lamps or automobile headlights; e.g. the nebulosity in the Pleiades, figure 2.2.

Other forms of interstellar matter are studied through the radio waves they emit:

Continuum Radio Emission (1932)

This can be compared to white light, and is radio energy emitted over a broad wave band. Interstellar space is filled with high-energy particles (mostly electrons and protons) moving nearly at the speed of light, which are called *cosmic rays*. When these cosmic ray electrons encounter interstellar magnetic fields the particles are forced to spiral about the magnetic lines of force. In that process they lose energy in the form of radio waves which range over much of the radio spectrum (chapter 17). This emission is concentrated in the Milky Way and is most intense from the direction of the galactic center.[12]

Emission from Interstellar Hydrogen (1951)

Neutral (un-ionized) hydrogen atoms can absorb or emit tiny amounts of radio energy at a wavelength of 21 centimeters (see chapter 17).

Emission from Molecules (Post 1963)

A large number of complex molecules have recently been discovered in interstellar space. This form of matter is observable because many molecules generate characteristic radio frequency spectral lines (see chapter 19).

Optical and radio data have provided most of the available information on interstellar matter, but balloon, rocket, and satellite astronomy in the infrared, gamma-ray, x-ray, and ultraviolet parts of the spectrum are of great importance.

Trumpler's Proof

It was Trumpler's work, however, that produced the most dramatic quantitative proof of the effect of interstellar matter on the light from stars in clusters. He directly demonstrated the effect on the apparent diameters of open clusters (figure 8.3). He began by assuming that the the diameters of a large set of clusters were very close in size. For the nearest clusters he was able to find a distance using Cepheid variable stars. If the distance is known, the *apparent* magnitude (how bright a star appears to the observer) of the cluster stars can be converted into a true or *absolute* magnitude. He then used the Hertzsprung-Russell diagram, which plots the apparent magnitude of the stars against their color (i.e., temperature). Most stars lie on a band across this diagram, the so-called main sequence, which shows that stars are very ordered in their properties. The stars in each cluster lay on the main sequence, but the data for the different clusters could only be made to overlay each other by shifting the apparent magnitude scales. Since he knew the distance to some of the clusters, this shift allowed him to derive the distance to the other clusters. Inherent in his work was the knowledge that the majority of stars of known distances always lay on the main sequence and what caused them to lie off it was only an incorrect estimate of their distances. (The relationship between apparent brightness and absolute brightness is, unfortunately,

FIGURE 8.3. Open star cluster NGC 2682 in Cancer. Trumpler's studies of the diameters of open clusters allowed him to prove that interstellar matter existed and that it was systematically dimming the light from more distant clusters. (Palomar Observatory photograph)

confusing to anyone not familiar with these concepts, but this summary is given anyway, for completeness.)

Now he had found the distance to all the clusters in his sample, but he noticed that the apparent diameter for the more distant clusters appeared consistently smaller than expected if they were all of the same typical diameter. In fact, the trend was so strong that he could not explain it except to infer that his original data had been in error. The brightness of the stars in the more distant clusters had been dimmed by passage through space—through interstellar matter, in fact.

Another way to state this, and as more commonly quoted in reports of this paper, is that if the apparent diameters of the clusters are used together with the distance estimates, the real diameters are found to be larger than expected and, furthermore, increase with distance. This is the consequence of having overestimated the actual distances. This, in turn, is expected if light is absorbed by intervening dust. Trumpler was able to show that the absorption amounted to about 0.7 magnitudes per kpc.[13]

This work was a triumph for astrophysics because it brought together so much information acquired for the study of stellar spectra, temperatures, and distances using Cepheid variable stars.

The study of selective absorption, i.e., changes in the color of stars, studied by Kapteyn in 1909, had also shown an effect which implied absorption of 0.3 magnitudes per kpc. In those measurements the color of the star indicates its temperature and hence spectral type. An independent observation of spectral type should lead to the same temperature, but did not, and the difference was taken to be due to absorption effects. But Shapley had found no color effect in globular clusters (since there is little or no obscuration well away from the galactic plane), while Trumpler found enough color excess in open clusters to indicate 0.3 magnitudes of absorption per kpc.

Trumpler pointed out that because the absorption showed a wavelength dependence, the data indicated that the particles involved had to be of a size comparable to the wavelength of the light. Following Eddington, the concept of meteoric particles was again invoked, although that label was subsequently to disappear from usage.

He summed up what the observations proved about the nature of interstellar matter. "It seems very probable that there is some relation between the medium causing our general and selective absorption on the one hand, and interstellar calcium and obscuration effects on the other hand."[14]

He concluded that several forms of matter had to be present in space. Atoms—calcium and sodium—produced the stationary lines. Free electrons had to be present since the calcium was ionized and the electrons, stripped off these atoms, had to be out there somewhere. Fine cosmic dust had to be exist in order to account for the selective absorption. Finally, large meteoric bodies should exist in order to account for the general absorption.

Today only the last conclusion would be brought into question unless we allow that there exist as yet unidentified large bodies in space (see chapter 24). The particles that absorbed the light need not be so large, however, because fine dust in large quantities, in extended but diffuse dust clouds, for example, would account for the general absorption. While these ingredients are usually mixed, some dominate in certain clouds, while other forms of interstellar matter, or other species of atoms (and molecules), dominate in other regions.

In 1937, Olin C. Wilson and Paul W. Merrill[15] presented an extended review of observations of interstellar sodium which also produced a "stationary" line. In retrospect, the work on calcium discovered in 1904 never did become the focus of attention. Also in 1937 the first major paper on the possible existence of interstellar molecules appeared. Pol Swings and L. Rosenfeld[16] suggested that searches be made for CH, OH, NH, and CN. CH and CN would be quite quickly discovered while the detection of OH had to wait until 1963.

By 1936 Otto Struve and Helen Story, in reviewing the evidence for the existence of interstellar matter, could write, "It is now generally believed that in addition to free atoms which give rise to 'stationary' absorption lines, interstellar space contains an appreciable amount of finely divided matter which causes selective, as well as nonselective, absorption."[17]

Eddington Again, and This Time He Sets Matters Straight

Eleven years after his important 1926 review Eddington provided an update of the subject of diffuse matter in space in a delightful essay in *Observatory Magazine.* This august magazine celebrated its sixtieth birthday by inviting famous astronomers to discuss current problems in astronomy. This time Eddington gave Hartmann his due as the discoverer of the interstellar calcium lines. The evidence that the stationary lines were produced in interstellar space had finally become overwhelming. These lines had previously been seen only in the spectra of the relatively younger, brighter stars, and the possibility that the gas was associated with those stars could not, at least initially, be ruled out. However, in 1926 J.S. Plaskett had shown, on the basis of data for forty stars, that the calcium was interstellar. Eddington had also rediscovered the work of Slipher, who had shown in 1909 that the calcium was interstellar, and noted that his work "has escaped attention."

The crucial piece of information that proved beyond any doubt the interstellar nature of the absorption lines came when Plaskett and Pearce showed that the velocities of the lines were, on average, half as great as that of the stars. This was expected if the gas lay between the stars and the sun.

Eddington also had to reexamine his former contention that the gas had a temperature of 10^4 K and a density of 10^{-24} gm/cm^3. Since ionized calcium and neutral sodium had been detected in this space, the temperature could not be that high. If it were, then twice-ionized calcium and ionized sodium would have been detected, and they weren't present. Thus Eddington confessed that, "I should ... accept without demur a substantial amendment of the above figure to suit the observed intensities of the fixed lines; but it seems impossible by any juggling to reduce appreciably the disparity in the proportion of Ca^+ (ionized calcium) and Na atoms."[18]

The relative abundances of sodium and calcium were difficult to understand. He said that "Presumably hydrogen is abundant in the cloud, as in other celestial objects.... In the nebulae the hydrogen gradually eliminates all radiation capable of ionizing it."

This was a wonderful example of his insight. Hydrogen was later discovered to be the principal constituent of interstellar matter. In the nebulae, the hydrogen atoms absorb ionizing radiation from newly formed stars to create incandescent ionized gas, which causes the Strömgren spheres around isolated stars and star clusters to ionize surrounding matter, producing an emission nebula. Eddington went on,

But the important conclusion is that a number of important elements H, O, N, etc, will be entirely un-ionized in the cloud. Ionized atoms repel one another and have no chance of forming molecules. ... the new conclusion [concerning the neutral atoms] seems to make it likely that the above mentioned elements will frequently form molecules. Also there is no obstacle to molecular combination of an ionized atom with a neutral atom.[19]

Eddington was now being far more accurate in his forecasts of what was to be discovered in the realm of interstellar chemistry than he had been eleven years before. Combinations of neutral atoms and ionized atoms are indeed common in space. He went so far as to suggest that a considerable part of the clouds might be molecular in form, a suggestion now borne out by the discovery of interstellar clouds, which consist entirely of molecules of one sort or another. He speculated about molecules banding together to form larger particles, *grains* as astronomers now call them, and pictured ice as existing in the clouds, something also borne out by subsequent observations. Recalling his first attempts to deal with the topic we must wonder whether this was the same man!

Atoms of iron and other elements would adhere to the ice-crystals and be dissolved by them, and we might conjecture that, on approaching the Sun, the water evaporates leaving a residue to enter our atmosphere as a meteor.[20]

Meteors are believed to have their origin in cometary material which, it is argued by some, are of interstellar origin (chapter 9). Comets are known to have the layers of water in the form of "dirty ice" mixed with solid material.

Eddington was in full cry.

Perhaps my inclination to a "water theory" is influenced by the feeling that, inasmuch as the gas responsible for the mysterious light of the nebulae has turned out to be just *common air* [referring to the presence of the elements H, O, N], it is fitting that the dark obscuring massed in the Galaxy should turn out to be just common clouds.[21]

Today this analogy would not stand closer scrutiny, although the image is not far off. The interstellar clouds contain mostly hydrogen, either atomic or molecular, and proportionately far less nitrogen or oxygen than is found on earth, yet at the time it was a marvellous way of bridging a conceptual gap. What are the clouds in interstellar space really like? Answer: They contain dust and gases and water (found in 1968) and marvellous molecules (all found since 1968).

Finally he asked, "I wonder what our successors will have made of this great problem when the centenary of the *Observatory* arrives?"[22]

The centenary of *Observatory* passed in 1977. It was unfortunate that the journal did not see fit to reinstitute articles on the same subjects as their fascinating series forty years before. By then dozens of interstellar molecules had been discovered and the nature of the mysterious light from the nebulae was well understood as being due to radiation from ionized hydrogen gas. What will our view of the interstellar matter, its origins and evolution, be a hundred years from now?

By 1930 the existence of interstellar gas was firmly established but the question of general absorption of starlight was still at issue, hence the size of the galaxy based on studies of the distribution of stars remained uncertain.[23]

A Summary of Interstellar Matter

Before we make the leap to the modern view of interstellar matter it is worth providing an overview of what has been discovered concerning the nature of matter between the stars.

Interstellar space is vast and at first appearance empty. However, it contains enormous diffuse clouds of matter consisting mainly of molecular and neutral hydrogen gas and a small proportion of heavier atoms and molecules such as calcium, sodium, water, ammonia, formaldehyde, and carbon monoxide. It also contains large quantities of microscopic solid dust particles of uncertain composition as well as magnetic fields which thread their way between the stars. The amount of interstellar matter in the Galaxy is about 10^{10} times that contained in the Sun (that is, about 5% of the Galaxy's total mass). Most of this is in the form of molecular hydrogen. Interstellar gas and dust generally occurs in cloud-like concentrations which may condense to form stars. Many stars, in turn, lose mass back into space in catastrophic explosions called supernovae or in less violent upheavals of material from planetary nebulae, novae, or, as recently discovered, in the constant ejection of material which flows away from newly formed stars. In these ways processed stellar material is returned to interstellar space where it is available for further star formation. Every atom in your body has been formed in a star and has been repeatedly cycled back and forth between stars and interstellar space before settling on earth during the formation of the solar system.

Interstellar matter is concentrated in the spiral arms in the Galaxy. This is dramatically illustrated in figure 1.2 (and also in figure 17.1 and 23.1) which reveal that interstellar dust clouds define the spiral arms of a galaxy. The visible Milky Way is a the disk of our galaxy seen edge-on, where the spiral arms lie hidden, one behind the other. Interstellar dust clouds close to the Sun are visible on a clear night in the northern summer as dark bands running through the constellation of Cygnus.

Interstellar matter is found in several distinct phases which have different densities, temperatures, and degrees of ionization:

FIGURE 8.4. The central regions of the beautiful emission nebula M16, revealing towering pinnacles of interstellar dust reaching across space between the stars. Luminous edges to the dark masses indicate surfaces where matter is literally boiling away. Intensely dark blobs are seen projected in front of the luminous background. These are examples of Bok globules, very dense structures within which stars may yet form, named after Bart J. Bok, who studied them extensively. The variety of dust masses in this image contain complex organic molecules as well as vast quantities of matter, which may form not only stars, but planets as well. This photograph also reveals how far the art of astronomical photography has been taken, in this case by David Malin using the 150-inch, compared to Barnard's pioneering efforts around the turn of the century with much smaller telescopes. Compare this photograph with Figure 21.5, an infrared image that penetrates the dust in M17. (Copyright Anglo-Australian Telescope Board)

1. A cold, neutral, phase in which atomic and molecular gas is concentrated in clouds from a few to hundreds of light years in size. The clouds have temperatures of a 100 K or less, and densities of from 1 to 10^6 atoms per cubic centimeter (cm^{-3}).
2. A hot intercloud phase, at temperatures around a few thousand K, density 10^{-2} cm^{-3}, which is widely spread and whose presence is inferred by its influence on radio waves from the Galaxy, pulsars, and distant radio sources.
3. A hot, ionized phase, known as the coronal gas, at a temperature of 10^5 to 10^6 K, density 10^{-3} cm^{-3}, distributed through much of interstellar space. This phase is observed through studying ultraviolet absorption of starlight by ionized interstellar oxygen as well as the x-rays emitted by the coronal gas.

Figure 8.4 illustrates several examples of interstellar matter. This is a photo of a section of the emission nebula known as Messier 16. It is produced by a cluster of very young hot stars whose ultraviolet radiation causes the surrounding gas to heat up to 10,000K. Surrounding the stars are interstellar gas and dark dust clouds several light years in size. The illumination and heating of these clouds produces the dramatic structures seen in the photograph. The intensely dark regions, where relatively few stars are visible, are produced by very small dust clouds, known as globules, which drift in front of the luminous gas. The larger, more diffuse dust clouds which surround such nebulae contain mostly molecular hydrogen and a large variety of interstellar molecules. The cluster of young stars that gave rise to this nebula was born out of the cloud which is now illuminated by these stars. Bright rims are seen where gas is being "boiled" from the surface of surrounding dust clouds.

The next question was: What was the nature of the material that caused the obscuration of light, that gave rise to Barnard's dark markings?

Notes

[1] Trumpler, R.J. (1930) "Absorption of Light in the Galactic System." *Publ. Astr. Soc. Pac.*, 42: 214.
[2] Shapley, H. (1917) "Studies Based on the Colors and Magnitudes in Stellar Clusters, First Part, The General Problem." *Astrophys. J.*, 45: 130.
[3] Curtis, H.D. (1917) "Absorption Effects in Spiral Nebulae." *Proc. Nat. Acad. Sci.*, USA, 3: 678.
[4] Seeley, D., Berendzen, R. (1972) "The development of research in interstellar absorption, c1900–1930." *J. Hist. of Astr.*, 3: 52.
[5] Globular clusters are closely packed agglomerations of old stars, sometimes thousands to a cluster, which are linked to our Milky Way by mutual gravitational attraction and drift in a spherical "halo" about the center of the Galaxy like flies around a picnic table.
[6] Shapley, H. (1918) "Globular Clusters and the Structure of the Galactic System." *Publ. Astr. Soc. Pac.*, 30: 50.

[7] Berendzen, R., Hart, R., Seeley, D. (1976) *Man Discovers the Galaxies*. New York. Science History Publications.

[8] Open clusters (once known as galactic clusters) are collections of from several dozen to several hundred relatively young stars born out of the same cloud of interstellar matter. Because of their youth (a few million to a few tens of millions of years old) they are located in the same regions of space where the greatest amount of interstellar matter is found, the plane of the Milky Way galaxy.

[9] Berendzen, R., Hart, R., Seeley, D., op. cit.

[10] Ibid., p. 77.

[11] Kerr, F.J., Henning, P.A. (1987) "Searching at 21 centimeters for galaxies behind the Milky Way." *Astrophys. J.*, 320: L99.

[12] See Gerrit L. Verschuur's *The Invisible Universe Revealed* for more details. Published by Springer-Verlag, New York, 1987.

[13] Trumpler, op. cit.

[14] Ibid.

[15] Wilson, O.C., Merrill, P.W. (1937) "Analysis of the intensities of the interstellar sodium lines." *Astrophys. J.*, 86: 44.

[16] Swing, P., Rosenfeld, L. (1937) "Considerations regarding interstellar molecules." *Astrophys. J.*, 86: 483.

[17] Struve, O., Story, H. (1936) "Scattering of Light in Diffuse Nebulae." *Astrophys. J.*, 84: 203.

[18] Eddington, A.S. (1937) "Interstellar Matter." *Observatory*, 60: 99.

[19] Ibid.

[20] Ibid.

[21] Ibid.

[22] Ibid.

[23] Berendzen, R., Hart, R., Seeley, D., op. cit.

CHAPTER 9

"Dark Nebulosity . . . Absorbs the Light"

Overlooked Voices on the Subject of the Dark

At last the notion of dark matter between the stars became widely accepted, but this happened only after the famous astrophysicist, Sir Arthur Eddington, turned his attention to the subject. It was his voice of authority that gave the stamp of approval and helped establish the widespread acceptance of the concept.

The words of the lesser known A.C. Ranyard, who in 1894 had admonished Barnard for believing in "holes in the heavens" and urged that the existence of dark matter between the stars be considered, continued to be ignored. But Ranyard was not the only astronomer who had pointed out the problem implied by the image of holes between the stars or had urged that the existence of interstellar matter be taken seriously.

Heber D. Curtis, whom we have met before, had in 1918 commented in regard to the dark markings seen projected against the bright nebula, M8 (figure 9.1). "I have never been able to believe that these are really "holes" in the nebula."[1] It was much more reasonable, he said, to suppose that they were masses of dark matter. Holes had to be shaped like tunnels all pointing away from our position in space, which was "against all probability."

It is impossible to look at the original negatives of these interesting objects and not be convinced that there is something dark between us and the general background of stars; I firmly believe that these are actual "dark nebulae."

He pointed out, as had others, that the random motions of stars should in any case be expected to fill in such holes, or at least blur their edges. This brought Curtis's tally of overlooked insights to three.

However, neither Curtis nor Ranyard was the first to give the correct explanation for the dark markings. Judging from the comments of astronomers in the early part of this century, none of them knew that there was an astronomer associated with the Vatican who had understood the correct explanation in the mid-nineteenth century.

FIGURE 9.1. The Lagoon nebula, M8, an emission nebula shining with the light from gas heated to incandescence by stars in a cluster deep within the nebula. Dark bands criss-cross the foreground and small globules, regions of incipient star formation, are seen in silhouette. The bright rims at the edge of several of the dust structures are produced by matter boiling away at the interface between cold and hot regions in the nebula. (Copyright Anglo-Australian Telescope Board)

Father Angelo Secchi—Shouldn't He Receive His Due?

In 1853 Father Secchi, of the Collegio Romano, announced in print in his local publication that

in these studies [of nebulae] one interesting fact stands out: namely the probable existence of dark masses scattered in space. These dark masses were seen because of the light background on which they are projected. Up to now these masses have been classified as "dark holes." This interpretation [of dark holes] is however very improbable, particularly after the discovery of the gaseous nature of the nebular masses. It is more likely that the darkness is the result of a dark nebulosity, which seen against a lighter background, absorbs the light.[2]

Who was this astronomer whose thoughts on the dark regions was unknown to the majority of U.S. astronomers in the early part of the twentieth century? As we explore that question, we cannot help but lament the fact that the Church did not see fit to oppose Secchi's views and thus give them the publicity which would have assured their rapid acceptance.

Secchi was born on 18 June, 1818 in Reggio nell'Emilia near Bologna. "His parents were honest folk of pious sentiments and highly respected by their fellow citizens."[3] (His father was a cabinetmaker.) He began his studies with the Jesuits in his hometown and then went to the Collegio Romano where he pursued his passion for science. When the Jesuit order was banished from Rome in 1848 he continued his studies first in England and then at Georgetown University in Washington, D.C. When the restrictions on the order were lifted the following year he returned to the Collegio Romano, where he began to revitalize its observatory and astronomical research program, putting the emphasis on astrophysics.

Secchi took some of the very first astronomical photographs when he made several daguerrotypes of the solar eclipse in 1851. He also became more interested in studying the physical nature of comets than in measuring their positions.

Secchi focussed his activities on the study of stellar spectra, a decidedly modern task. He was one of the pioneers in the subject, and Secchi's system for classifying stars according to the appearance of their spectra was widely used. He found that there were relatively few distinct-looking stellar spectra to be found among thousands of stars and his types became the basis for the classification system.

This focus on what were the beginnings of astrophysics did not go unnoticed, as Secchi himself noted concerning the choices made after he took charge of the observatory.

We, after several months of endeavor, concluded that research in precise astronomy was a waste of time and energy; so we had to limit ourselves mostly to comet research, and to studies of the physical appearance of the sun and other celestial bodies.

This concept was one that governed all our later research and the choice of instrument for the new observatory. According to some this was going too far and it was even said at the Collegio Romano that the science of physics rather than that of astronomy was being cultivated.[4]

The mention of physics thus sounded like taboo, but nevertheless he was able to work on, unhindered by dogma.

It was Secchi who, in 1859, coined the word *canali* to describe the markings on Mars which later evolved, with the help of other astronomers, into the misconception that there were water-filled canals on that planet. He also observed the nebulae and classed them as planetary, irregular, or elliptical. His work on the spectra of the bright nebulae revealed that they were hot, gaseous bodies, as Herschel had already suspected.

Forty years before Barnard had begun to photograph the dark markings (figure 9.2), Secchi had suspected that they presented evidence for obscuring matter over the sky as well as in some of the "elliptical" nebulae, which we can safely infer referred to distant galaxies seen edge-on. Secchi was aware of Herschel's belief in vacancies in space and disagreed with that notion. To him the explanation seemed

... quite improbable, especially after the discovery of the gaseous nature of the nebular areas and it is instead more probable that this blackness results from a dark nebulosity projected on a lucid background and intercepting its rays.[5]

He also made the connection between these dark markings and those that could be seen in the Andromeda nebula, a very perceptive and correct conclusion.

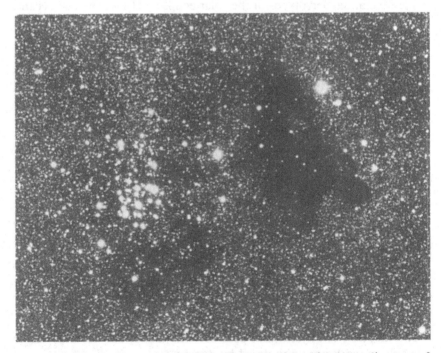

FIGURE 9.2. The dark marking known as Barnard #86, which has a diameter of 5 arcminutes. This modern photograph gives a sense of why it was possible to consider that these dark marking might be "holes in the heavens." A small cluster of stars is visible to the east (left) of the nebula. In this photograph millions of galactic stars are visible, and one has a sense that the region outside the dark marking has a glow of its own. (Copyright Anglo-Australian Telescope Board)

In 1873 the Jesuits were once again expelled from the Collegio Romano, but Secchi was allowed to remain at the observatory through the intervention of the government.[6] He died in 1878. At the church of San Ignatius the urn that holds his ashes was adorned with a red ribbon that carried the gold medal given to him by Napoleon III at the Paris Exposition in 1867. Secchi was a prodigious researcher who believed that "contemplation of the heavens is a sure way to God."

Why Are Some Voices Ignored?

By now it is obvious that several astronomers had the right idea about the dark markings that Barnard spent so much time photographing. The correct interpretations were aired at least eighty years before their final acceptance. Why did it take so long?

Here we must digress for a moment on how science, and in this case astronomy, is done. One the one hand the notion of dark matter between the stars was a radical departure from previous beliefs. But it was not the first time that some new idea had been proposed which flew in the face of long-held dogma. The usual pattern of events is then that others join in the quest to understand whether the new idea carries weight.

A second factor comes into play, and that concerns who makes the suggestion. The new idea must not only come at the right time, but it has to be uttered by the right person. The budding astronomer who reads this book may well bear this in mind. Sometimes a potent idea is proposed by someone with no reputation of the appropriate sort, and it will be lost. When an astronomer with "clout," a world figure such as an Eddington, makes the pronouncement, everyone listens. In this case they will listen even if he utters nonsense, in the hope that gems will be panned out of the grit of his statements.

Let us now look briefly at the various scientists who had pronounced correctly on the nature of the dark nebulae.

Secchi was known for his work on stellar spectra, but then only in a very limited sense. He was far away, relatively hidden from the "mainstream" of astronomy, as it were. His comments on obscuring masses between the stars were published in a place where it is unlikely that any prominent astronomers in the United States or even other parts of Europe ever saw it.

Ranyard was editor of a popular astronomy magazine and secretary of the Royal Astronomical Society, neither role being one which would cause the "establishment" to take him seriously, should it come to deciding a major research problem one way or the other.

Barnard was a self-made astronomer who had managed to join the ranks without attending a major academic establishment. His colleagues could not ignore his observational contributions, but one suspects that his effort to present the world with views of the Milky Way and its dark markings were widely regarded with benevolent patience. This would help explain why no others joined him in what surely would have been a valuable research project, to understand the nature of these markings.

The words of Curtis were overlooked because he was running headlong into the wall that surrounds any idea that cannot be proven in its day. It is perhaps just coincidental that Curtis began his academic career as a Latin and Greek scholar and graduated with little scientific, and no astronomical, training. He felt the lure of astronomy soon afterward and by age thirty had obtained a Ph.D. from the University of Virginia.

The key issue was that none of the protagonists for the idea were initially able to prove anything; they could only suggest, and although their suggestions were intuitively sensible, no one else took up the cause of seeking proof. Except, of course, Max Wolf, who was able to demonstrate that obscuring matter in diffuse regions seemed to exist.

What is extraordinary is that despite the lack of proof for obscuring matter, the alternative picture of holes in the heavens required even greater "belief." Yet this concept remained in place even if, as Ranyard and Curtis had pointed out, it flew in the face of common sense. Instead it took Eddington's authority to draw attention to the topic at all. (Several representatives of an older generation of astronomers, who have done important work on interstellar matter, told me that Eddington's review formed the center piece for their learning and interest in the subject.)

And Finally Some Comments on How It Is Said

Astronomers, like most other professionals, are jealous of their terrain. Thus when a Ranyard or a Barnard, outsiders as far as the establishment was concerned, utter a profound truth they go unheard. Had William Herschel, the musician from Germany, not discovered Uranus it is extremely unlikely that he would have been heard either, an exception that proves the rule. It is important to have risen through the ranks in the appropriate way in order to be heard, and even then it requires the backing of an authority figure to lend credence to one's ideas.

Today an Edward Barnard would have no chance of making it as a professional astronomer. A Ranyard would still not be heeded. A William Herschel, a foreign-born musician aged thirty-six seeking entrance to the system would not be given the time of day by the establishment. Does this mean we are all a little poorer for it, or that we can rest assured that the standards of the profession are higher?.

A side effect of the perpetuation of the elite is an extraordinary fragmentation of the community of citizens with an astronomical interest. Three groups with a deep and abiding interest in astronomy exist in the United States. The professionals, amateurs, and planetarians (those associated in some way with planetarium work) with few exceptions hardly speak to one another. The most dramatic aspect of this split is that a great deal of direct communication about astronomy to the public occurs almost exclusively through planetariums. The professionals remain essentially isolated from the other groups. At the same time the amateurs infiltrate the planetarium community and, by definition, hardly contact the professional community. It is nevertheless possible for an amateur to make an astronomical discovery, however rare it may be, most often that of a comet, nova, or supernovae. Intelligent, creative individuals in the amateur or planetarium community who have a profound insight worthy of further study will find that there is virtually no way for their ideas to be considered by the professionals.

These comments have a bearing on the broader issue of how astronomy is communicated, both to the public and to the professional. Innate elitism prevents the professionals from taking seriously any idea which does not have the correct label attached to it, a label that identifies who the idea came from, how it was presented, and where the proposer works or was educated.

Barnard's background did not sit well, even in his day, and he was aware of that. He did have the advantage that he was able to bring to astronomy unique talents, much needed when photography began to be exploited for research. The early twentieth century was also more forgiving of one's background, as Milton Humason's example reminds us. He was the mule driver who carried components for the Palomar Observatory up the mountain and later began to work at the observatory, and rose to making original observations so that his name came to be associated with the early work on the expansion of the universe.

And so it goes. Every individual has to make his or her own way through the system, but the system is not structured to encourage or help "outsiders." Their fight has to be more tenacious. Looking about at the astronomical scene today, with its far-reaching network of academic institutions churning out bright astronomers, the opportunities for a Barnard, a Herschel, or a Struve are few indeed, and very much more daunting than they were a half century ago.

Notes

[1] Curtis, H.D. (1918) "Dark Nebulae." *Publ. Astr. Soc. Pac., 30:65.*

[2] Abetti, G., Hack, M. (1959) *Nebulae and Galaxies*, New York: Thomas Y. Crowell, p. 20, from the original "Astronomia in Roma nel pontificate di Pio. IX—Secchi."

[3] Abetti, G. (1960) "Father Angelo Secchi A Noble Pioneer in Astrophysics." Leaflet No. 368, *Astron. Soc. Pac.*

[4] Ibid.

[5] Ibid.

[6] *Dictionary of Scientific Biography.* (1980) Charles Scribner and Sons, New York.

CHAPTER 10

The Nature of the Dust

Interstellar Dust

What is the composition of the dark clouds that gives rise to the obscuration of starlight? Just before Barnard died in 1923, Henry Norris Russell was the first to suggest the answer. He accepted the idea that the dark markings were due to obscuring material and said that in order to explain the observations this matter had to be in the form of fine dust.[1] Today astrochemists study the composition of fine dust particles in laboratory experiments meant to simulate conditions in the dark nebulae.

Russell realized that because the dark clouds had clearly delineated shapes something had to be acting to hold them together. Gravity was the only obvious candidate that could provide the force, and this assumption allowed certain cloud properties to be derived. Based on estimates of the size of clouds seen projected against nebulae at a known distance, Russell calculated the amount of mass involved. He was stunned to discover that it was enough to form many additional stars but the size of the particles had to be less than 0.1 mm in diameter in order to account for the way light was absorbed. He also explained the relationship between the dark nebulae and the two types of bright nebulae. Starlight bouncing off the dust particles would produce *reflection nebulae* of the kind seen in the Pleiades (figure 5.1), while others would shine if the gases that coexisted with the dust were heated to incandescence by radiation from hot stars. The great nebula in Orion (figure 1.5) is such an *emission nebula*. Russell did not know what radiation this might be and referred to it as some sort of "ethereal or corpuscular" emanation. The source of heat turned out to be ultraviolet radiation from hot stars.

In one brief paper Russell had succeeded in relating the various types of nebulae in interstellar space—the dark and bright ones—and proposed the existence of interstellar dust. His views remain essentially correct today, although it required Eddington's broader forum to spread the word.

Particles of cosmic dust are indeed very small and have an elaborate core-mantle construction, as inferred from measurements of the way they extinguish light, ultraviolet, and infrared passing through interstellar clouds. The dust grains are really huge molecules that made of millions of atoms. These grains may control the processes of star formation which, in turn, has a bearing on our existence on the planet Earth as conscious entities capable of comprehending the wonders of space.

The Life Cycle of Interstellar Matter

Interstellar dust grains exist in vast clouds throughout the Milky Way. In smaller quantities these grains exist everywhere between the stars. Sometimes the clouds gather to form larger aggregates which may collapse under the action of gravity to form stars, usually in groups known as clusters (see figure 8.4). Following the stage of star formation, some of the gas and dust is ejected back into space where it rejoins diffuse clouds. At the end of a star's life a second phase of mass ejection, sometimes due to a violent stellar explosion, injects processed stellar gas back into space. Figure 10.1 summarizes the picture.

In the early universe all matter was hydrogen and helium gas created in the big bang and thus interstellar matter contained little more than

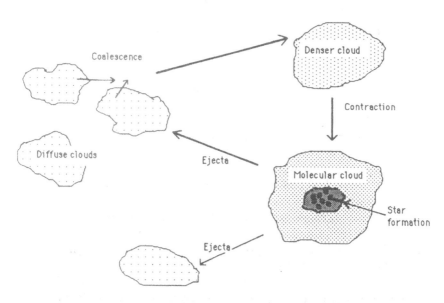

FIGURE 10.1. The life cycle of interstellar clouds. Diffuse clouds coalesce to form denser clouds which may reach the point of contracting gravitationally. Molecule formation occurs on the surface of the dust grains, and if the density of the molecular cloud becomes sufficiently large, clusters of stars form. The energy produced by these stars is sufficient to disrupt what is left of the original molecular cloud, which fragments. The ejecta, no longer controlled by gravity, expand and become diffuse clouds in search of new partners with which they may again coalesce.

those two elements. It would be constantly stirred, mixed, and shaped into stars where the gas was cooked in stellar ovens, percolated to the surface, and evaporated back into space through ejection of material in stellar winds. As eons pass, interstellar matter changes its chemical composition, inevitably evolving a higher fraction of heavy elements. These "metals" subsequently form the matrix of solid dust grains which obscure starlight. The grains are also raw material for planetary formation.

The sun and earth were formed from interstellar material, both the gas and dust. The earth consists of highly condensed, differentiated interstellar matter which has been further processed in the earth's core, on its surface, and in the oceans and atmosphere. The core of the earth contains mostly heavy iron, the outer layers contain lighter elements, while the atmosphere consists of light gases. Five billion years ago all these materials drifted together in interstellar space.

Interstellar matter continually rains down on our planet. Comets, for example, may be the most direct source of reliable information as to the original state of the interstellar dust, this according to J. Mayo Greenberg of the Astrophysical Laboratory of the University of Leiden Observatory, who has been studying the nature of the mantles of interstellar grains in the laboratory. He believes that comets are the relics of the material that 4.7 billion years ago formed the solar system and thus carry within them traces of raw interstellar matter.

An Interstellar Grain Revealed

Figure 10.2 is a highly enlarged view of what an interstellar dust grain may look like, according to Bertram Donn and Joe Nuth at the Goddard Space Flight Center in Maryland.[2] This particle, produced in laboratory simulations, has a "fractal" form, which refers to the detailed structure of the way individual bits are connected in a pattern which repeats itself on ever smaller scales.

In 1935 the Swedish astronomer Bertil Lindblad had suggested that grains might be formed by condensation processes, and in 1949 Henk van de Hulst in Holland proposed that the condensates might be ices. But it was also realized that such condensation required nucleation sites, much as water vapor in terrestrial clouds requires nuclei around which to condense in order to form raindrops. There appeared to be no way to form the initial nucleation sites in space until, in 1970, it was found that silicate particles surround cool stars called M giants and supergiants. These particles condense in the stellar atmospheres and are then driven into space. As soon as they move far enough away they cool to 10K and act as condensation nuclei to which other atoms and molecules stick. These migrate over the surface, mix, and combine to form a mantle containing ices and organic matter.

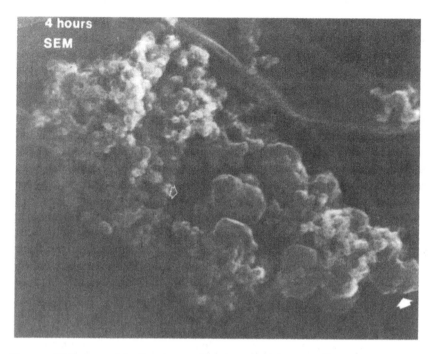

FIGURE 10.2. A model of an interstellar dust particle, according to F.J.M. Reit-meyer, J.A. Nuth, and I.D.R. MacKinnon. It has a fractal nature. (J. Nuth, NASA Goddard Space Flight Center)

Greenberg and his colleagues attempted to predict mantle composition by beginning with the most abundant elements—C, O, N, and H. These readily combine to form water, methane and ammonia. Greenberg calls this "dirty ice." Important chemistry results when these molecules break up and the pieces recombine to form more complex ones. Like corporate giants selling off subsidiaries, the fragments are restructured to form new entities with lives of their own.

The frozen layers of material in the dust grain are bathed by ultraviolet light from stars, intense in some regions of space near newly formed stars, and much less intense further away. Molecules are extremely susceptible to destruction by the ultraviolet light, undergoing *photolysis* , or decomposition by action of radiant energy. They may be broken up in a few hundred years but the grain cores last a long time, so the chemical changes on their surfaces are continuous as new bits and pieces stick, combine, and are again broken apart. Chemically very reactive radicals, such as OH and NH_2, are formed in abundance and help form new molecular species. The grain mantle consists of a hodgepodge of stuff, which no doubt accounts for the existence of over seventy molecular species in the gaseous state that have been observed in space. They must have been driven off the grain surfaces.

The core of an interstellar grain is believed to consists of silicate material mixed with other heavy elements, while the mantle is made of ices of various sorts deposited on the metal core. Donn and Nuth have been doing experiments meant to produce dust-like particles in conditions which simulate the atmospheres of stars, where the grains are believed to be formed. Greenberg and his colleagues attempt to simulate what happens when ultraviolet light shines onto the ices that condense on the grain surfaces in interstellar clouds, while Donn and his colleagues have performed similar experiments to study how particle radiation (cosmic rays) affects the chemistry on the grain surfaces (see chapter 20).

The laboratory experiments are meant to produce data that may be compared directly with astronomical observations. The way light from stars is dimmed by intervening dust clouds gives the absorption spectrum of the dust, as shown in figure 10.3. This is called the extinction curve. Details of such curves vary with the direction of the observation, because

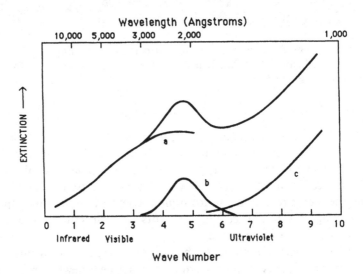

FIGURE 10.3. The shape of the "extinction curve" produced by interstellar dust clouds. The reduction or extinction of the amount of the light and infrared radiation passing through the cloud is indicated by the amplitude of the curve at the relevant wavelength. The shape of the curve is broadly approximated by invoking three separate regimes produced by particles of different sizes and compositions. Observations of extinction toward specific stars reveal differences in detail, accounted for by differences in chemical and physical properties of grains, related to local conditions, history of the grain, and so on. The horizontal axis is usually displayed in units known as wave number, which are an inverse of wavelength.

the nature of the interstellar grains varies from point to point in space, depending on the recent history of events in that specific corner of the galaxy.

The extinction curve is determined by measurements at optical, infrared and ultraviolet wavelengths. Dust does not influence radio waves which are far longer than the size of the dust particles, therefore the dust ignores them. The spectrum (figure 10.4) is complex and contains several important features that tell a great deal about the nature of the particles. A basic rule in interpreting such curves is that the particle size must be of the same order as the wavelength for the radiation being absorbed. Thus the extinction curve gives information about particle sizes.

In the infrared and visible parts of the spectrum (to the left in figure 10.3) the extinction is produced by grains about $1/10$ of a micron across (10^{-5} cm), the size of cigarette smoke particles. The ultraviolet extinction (on the right of figure 10.3) is due to particles $1/10$ as small (10^{-6} cm), consisting of silicates. The hump at 2,200 angstroms is due to equally small particles believed to consist mostly of carbon, perhaps in the form of graphite. But how can the grains have two sizes? Because of the agglomeration of small pieces into larger masses. The structural differences produce a range of extinctions across the observed wavelength domain.

Research on the chemical nature of the grains has been extensive and dozens of astronomers have contributed to forming the pictures described here. We can do no more than summarize some of the aspects. Two important constituents appear to be present; silicates in the cores and water in the form of ice in the mantle. Figure 10.4 is a schematic representation of the nature of an interstellar grain. Silicate grains are directly observed around older stars close to the end of their lives and there is now little doubt that the grains are formed in the stellar atmospheres and then ejected into space.

The silicate cores of the grains form nucleation sites on which other molecules, in particular organics (see chapter 20), may condense. The cores (figure 10.4) are quickly surrounded by a mantle of complex organic materials ("yellow stuff"; see below) and a further mantle of water ice. It is in this layer that the interesting chemistry takes place. The ice layer contains many elements and molecules. When energy in the form of ultraviolet radiation is introduced, from a nearby star for example, chemical changes occur which form more complex molecules. Through grain collisions or more direct heating these molecules may, in turn, be released into space where they are then observed as an independent species by radio astronomical observations.

The basic and important combination of two hydrogen atoms to form a molecule also occurs on grain surfaces because the hydrogen atoms cannot combine in the gaseous phase. This requires the presence of a third body, the grain, to dissipate the heat produced by combination.

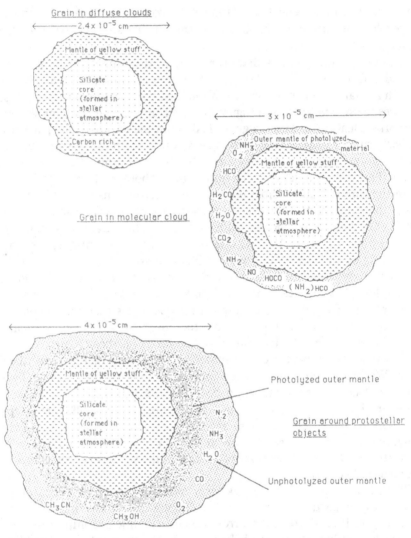

FIGURE 10.4. Model of the evolution of an interstellar dust grain, according to Greenberg and coworkers. Initially the grains created in the atmosphere of a cool star will consist only of the silicate core. The grain in diffuse clouds consists of this core surrounded by "yellow stuff" (see text), which presupposes that the grain has already gone through a cycle of cloud formation and destruction. In a molecular cloud, an outer ice layer is contaminated by a wide variety of molecules, which are further processed by photolysis (the incorporation of energy from radiation such as light). If the cloud reaches sufficient density to form stars, the grains around the protostellar object are further layered by molecules, which adhere to the grain.

Simulation of Interstellar Processes in the Laboratory

This simplified picture of grain chemistry can be simulated in the laboratory. Among the first to attempt this were Carl Sagan and B. N. Khare at Cornell, and Greenberg and Andrew Yencha at the State University of New York at Albany. Greenberg has pursued this work at the University of Leiden.

In laboratory experiments it is impossible to precisely simulate conditions in space, because the best vacuums on earth do not remotely approach those in space. Nevertheless, Greenberg uses the best possible vacuum and a "cold finger," a tiny block of metal, mounted in an enclosed container kept at 10 K, to simulate the core of the grain. His challenge was to perform chemical experiments similar to those occurring in space. Stars fill space with ultraviolet light, so the laboratory equipment has a powerful ultraviolet source that can shine into the container and illuminate the cold finger. Through other ports, infrared or visual light is beamed past the finger and into a spectrometer. This allows the absorption spectrum produced by the material on the finger to be analyzed to reveal chemical composition.

Gases are then introduced into the container; these include CH_4, CO, H_2O, CO_2, NH_3, N_2, and O_2. The ultraviolet source is then switched on and in an hour of laboratory time the equivalent of one thousand years of irradiation by ultraviolet light in diffuse interstellar clouds is simulated. An hour in the laboratory is also equivalent to as much as ten thousand to ten million years in dense molecular clouds (chapter 21).

Formaldehyde (H_2CO) and HCO, both common interstellar molecules, are quickly formed on the finger. Other organic molecules are created and, together with the ices, produce an extinction curves similar those found in the direction of stars. Water ice formed at 10K produces some of the detailed characteristics of the extinction curve, although the fine details are not shown in figure 10.3.

The chemical changes in the layer of ices surrounding the cold finger are profound and result in the production of complex molecules whose action closely mimics the observed interstellar extinction curve. But these results came not without an astounding discovery. At some point the material explodes.

When radicals come in contact with one another they release energy. Greenberg estimated that if a thousandth of the radicals on the grain all happen to combine at the same time they will raise the temperature by 20K. This frees other radicals, which then combine rapidly. The extreme heat melts more of the ices, allows more reactions, and in a matter of moments the grain literally explodes because of the energy liberated. Since the real interstellar grains are usually around 10K this is not a usual occurrence. Hence chemical changes may proceed at a gentle pace and will make ever more complex molecules. It requires heating to encourage the process which then releases molecules into space and stops the process.

Experiments in which grains with no water ice were allowed to warm slowly produced a bluish-green light from energy released by the combination of HCO radicals at their carbon atoms to form glyoxal molecules. More rapid heating generated more rapid combination of radicals and caused the sample to explode. The main explosion occurred at around 25K, and actually left the experimental cold finger bare of material.

When the experiment was gently ended and the sample allowed to reach room temperature, avoiding sudden explosions, a chemical residue was left behind which is quite stable. This organic goo, "yellow stuff" as Greenberg's team called it, is a carbon-based material of unknown composition that remains difficult to analyze.

These experiments are believed to produce the same mixture of molecules found in interstellar grains. The residues and the identifiable molecules, including amino acids, are similar to those produced in laboratory experiments that synthesize conditions on the early earth which led to the formation of life. In 1953 Stanley L. Miller and Harold C. Urey mixed similar gases in flasks and injected energy in the form of ultraviolet light and electrical sparks and found that amino acids as well as a host of other molecules, as well as the yellow stuff, were produced.

The birth of interstellar dust grains clearly requires nucleation cores which allow them to collide and grow in size. Only where the densities are high enough, millions of times greater than found in normal interstellar space, is their formation possible. These silicate cores are found in the atmospheres of cool, giant, oxygen-rich stars. Once formed, the grain cores are blown into space. The grain cores *cannot* be formed in space, but are born as seedlings ejected from old stars. The cores then undergo extensive chemical changes in space before being taken up in new stars. The amount of such material ejected into space in the process about balances the amount used up in star formation per year in the Galaxy, thus maintaining a continuous cycling of matter from space to stars and back into space again.

Grain Life Cycles

The average density of matter between the stars is now extremely small. One would need to search a cube whose size is the length of a football field, that is, a volume equal to the interior of the Houston Astrodome, to find one grain. Nevertheless interstellar clouds are so vast that even a small cloud contains enough grains to absorb starlight much as smoke in our skies absorbs sunlight. If the solar system were located inside a dense dust cloud all the stars would disappear from view, but the sky would be bright due to scattered light from the sun.

Once the grain finds itself in a diffuse interstellar cloud, it accumulates a mantle of ice. Ultraviolet light creates chemical changes and in due course it develops an inner mantle which may consist of yellow stuff, and an outer mantle with a substantial fraction of water. This is revealed by the presence of a three-micron[3] (3 μm) absorption band in the infrared. An increase in grain size produced by the growth of the outer mantle should produce some correlation between visual extinction and the strength of the three micron band. In some cases this is seen. Young stellar objects that emit infrared radiation, such as those known as Becklin-Neugebauer objects in the Orion nebula, contain grains 60% of whose outer mantles are water.

The grain doesn't grow indefinitely. The miniature explosions control growth, but these won't happen if the grains stay very cold. Collisions between grains can create warming, which may lead to explosive heating, which then feeds material back into space, where it is free to join another grain. And so the process continues, endlessly.

According to Greenberg, if one grain out of ten collides in a million years it is enough to maintain the density of interstellar molecules observed in the gaseous state (chapter 19). The grain may undergo ten catastrophic explosions on its surface in its lifetime in a cloud, and many gentle warming-ups which cause it to develop a fairly complex and stable inner mantle of yellow stuff.

The evolution of an interstellar cloud determines the ultimate fate of the grains. Some clouds coalesce and begin to shrink under the influence of gravity to forms stars. In the process most of the grains are destroyed and others are ejected back into space where they continue their lives. The grains that return to diffuse clouds may in due course end up in space between the spiral arms of the Galaxy (chapter 17) and lose volatile material from their surfaces. Collisions and supernovae further damage the grains. The thickness of the mantle of yellow stuff may then decrease.

In high density regions of molecular clouds (see chapter 21) all matter except H and He may be trapped in grains. The grains that then fall into a star in the process of formation are destroyed while others may escape. Some may be heated close to the star and evaporate while the lucky ones persist and produce some of the spectral features observed in the direction of such protostars (chapter 21).

In a grain's life destruction and formation processes occur for about equal times, so they spend as much time in diffuse clouds as in molecular clouds. Interstellar matter is therefore well mixed. The average composition is about the same everywhere. Greenberg estimates that the interstellar medium is consumed once in 5×10^9 years. Clouds last up to 10^8 years. Grains can pass in and out of dense clouds twenty-five times or more before ending up in a star, or a planet.

Comets—Packages of Interstellar Matter

Nature may have provided us with a regular shower of interstellar matter in a variety of forms. According to Greenberg and other astronomers, comets are enormous masses of interstellar grains. The formation of comets may come from the initial gathering of clumps of grains, then clumps of clumps, and so on. These cosmic visitors contain massive quantities of water with a mixture of organic molecules which may once have been as fluffy as freshly fallen snow. That snow now contains interstellar dust, which led Fred L. Whipple to describe them as "dirty snowballs." Greenberg thinks that the comet nucleus consists of 30% water, 20% yellow stuff, 8% silicates, 3% carbon, and 40% other molecules, which makes them "very dirty snowballs."

About 10^{11} to 10^{12} comets are believed to exist in the Oort cloud, a spherical halo of matter around the sun at distances between 20,000 and 100,000 times the sun to earth distance, or about 0.1 to 0.5 pc. These comets skulk about at interstellar distances, and when close encounters with other stars perturb a comet's movement it may wander close to the sun and earth and become visible from earth. Their cold nuclei are storehouses of interstellar matter and their well-known tails are produced when this matter boils off as the result of heating by solar radiation.

Michael F. A'Hearn at the University of Maryland and Paul D. Feldman at Johns Hopkins University recently found sulfur molecules in comets. These could only have been produced in solid matter bathed in ultraviolet radiation. Laboratory simulation has shown that S_2 can only be formed by irradiation of dirty ices containing H_2S at 10K to 15K. If the temperature goes above 30K, it is destroyed. Therefore, they concluded that the comets had to be made of virginal interstellar dust that was never heated above 60K. The S_2 is found close to the nucleus of the comet, where it has been untouched since formation. The very fact that it has been observed at all means that it has been protected since the comet formed; that is, since before the solar system formed.

Interstellar Matter on Earth

Meteors, especially those occurring in showers known to be related to old cometary orbits, are comet debris, so they are also forms of interstellar dust. Every day the earth is showered by meteoric material, which is sometimes seen as bright tracers across the night sky (shooting stars). Most of this meteoric dust settles unseen onto our planet, depositing 10,000 tons every day. Meteorites, the larger objects that strike the earth, appear to have different orbits which associate them with the asteroid belt, but we cannot rule out that some of those larger objects that have struck the earth recently have come from interstellar space.

Both meteorites and interplanetary dust particles gathered in balloon-borne experiments high in the atmosphere have recently been shown to contain a form of material known as polycyclic aromatic hydrocarbons (PAHs). These complex molecules produce excess infrared emission which is very similar to that seen in the interstellar extinction curve, and the PAH samples that have been analyzed contain more deuterium than is found in terrestrial examples of these molecules. Thus the PAHs must also have been formed in circumstellar dust shells and these have suffered only small changes before turning up on the earth.[4]

Black dust found in the Greenland ice cap has recently been shown to be extraterrestrial in origin and it too has remained unaffected by entry into the atmosphere. Chemical analysis proves that the dust was not produced by the breakup of larger particles. Some of the dust may have come from short-period comets, but there is too much to be accounted for in this way. A fraction must have come from collisions between earth-crossing asteroids and asteroid fragments. The dust created drifts onto the planet, where it remains undisturbed only in the ice caps. While it is certain that this dust comes from space, how deep in space is open to question. For example, it is possible that certain asteroids, especially the earth-crossing ones, were once comets.

Analysis of larger objects that strike the earth, such as the Murchison meteorite which landed in Australia in 1969, has revealed many organic compounds including amino acids containing high proportions of deuterium, which is rare on the earth but more common in space. The meteorite is therefore made up of material which existed at the time of the formation of the solar system and has been preserved unchanged ever since. Hence it is a direct sample of a large accumulation of interstellar matter. Just how many objects this size exist in the Galaxy? No one has yet tried to estimate it.

Microscopic diamond dust has also been found in several types of meteorites. The diamonds are extremely small, containing about a million carbon atoms and forming crystals 50 angstroms (5×10^{-7} cm) across. This is similar to the size of interstellar grains discussed earlier. The diamonds contain xenon and a proportion of isotopes, which make it impossible that they were formed in the passage through the atmosphere. They, too, have been formed in interstellar space well before the Solar System was born.

Greenberg has gone so far as to suggest that a substantial fraction of the Earth's water as well as its surface organic material has arrived via comets. In the early Solar System comets would have been abundant and repeatedly struck the planets and the moon. One good sized comet can supply 1.5×10^{17} gm of water. The moon's gravity was too low to keep the water, but it would have remained on the earth. The water contained in the oceans at present is contained in the equivalent of 10^7 10-km–diameter comets, not an unlikely guest list for a "planet-warming" party soon after the formation of the Solar System. At the same time, these early comets could have contributed 10^6 times as much organic molecules as are now found on earth. If only a tiny fraction of this interstellar matter survived impact, in the form of amino acids, say, it would have given a tremendous boost to the formation of terrestrial life. Even using more conservative estimates of how often comets strike the earth, it is quite easy to introduce organic molecules of the order of the earth's biomass onto the earth during the first 500 million years of its life. The organic molecules formed in space may, in fact, have provided the templates for the origin of life (chapter 23).

This idea for the origin of the terrestrial oceans has been dramatically strengthened by Christopher Chyba of Cornell University, who studied the distribution of craters on the moon to derive an estimate of how many large objects would have been in the vicinity of the earth soon after the time of its formation.[5] He found that if only 10% of these objects were comets, the oceans of the earth would have been formed by them in a period of only 700 million years. He assumed that comets consist of 50% water ice.

How impossible it would have been for Barnard to conjecture that the dark markings revealed in his photographs were clouds containing small solid particles layered in ices, samples of which rain on our planet, and may have formed the oceans and carried the seeds of life.

Notes

[1] Russell, H.R. (1922) "Dark Nebulae." *Proc. Nat. Acad. Sciences,* 8:115.

[2] Nuth, J.A., Donn, B. (1982) "Experimentals tudies of the vapor phase nucleation of refractory compounds." *J. Chem. Phys.,* 77:2639.

[3] The unit of wavelength used by infrared astronomers. One micron (μm) $= 10^{-6}$ meters.

[4] Allamandola, L.J., Sandford, S.A., Wopenka, W. (1987) "Interstellar polycyclic aromatic hydrocarbons and carbon in interplanetary dust particles and meteorites." *Science,* 237:56.

[5] Chyba, C.F. (1987) "The cometary contribution to the oceans of primitive Earth." *Nature,* 330:632.

CHAPTER 11

Odds, Ends, and IRAS

A Galactic Light Show

In February of 1901 a new star appeared in the constellation Perseus. It was a nova—not the violent destruction of a star, but a much less, albeit still catastrophic, shrugging off of a shell of matter which suddenly caused the star to flare in brightness. Several months later, in August, a remarkable nebulosity was found around Nova Persei. In the time interval between the two photographs shown in figures 11.1 and 11.2 ,the nebula changed its shape in such a way that it appeared to be expanding very rapidly. Had these photographs been fully understood in 1901 the question of the existence of dark interstellar matter would not have been in doubt for very long.

Figure 11.1 shows the nebula as it was on September 20, 1901, while figure 11.2 shows the photo taken on November 13, 1901.[1] A great ring of patchy luminosity appears to be moving away from the central star. If the distance of the nebula is known, the velocity of expansion can be calculated. At the time, Kapteyn reversed the argument and, by (incorrectly) assuming the expansion was at the speed of light, he found the distance to Nova Persei to be 90 pc.

Even as Kapteyn was drawing his authoritative conclusion about the distance to the nova, Arthur R. Hinks of Cambridge Observatory in England pointed out that if the effect was due to the lighting up of a surrounding nebula by a burst of light from the nova, one might expect to see apparent motions outward which should approach or even surpass the speed of light, depending on the way structure in the nebula is oriented with respect to the observer.[2] His suggestion was ignored.

In 1903 C.D. Perrine obtained a spectrum of a segment of the nebulosity (by now the nova had faded) and found it to contain precisely the same lines as were present in the original nova. This proved that the nebulosity was a light echo, a reflection from surrounding matter whose nature was, of course, not yet understood. He had demonstrated that it was light reflected off interstellar dust that produced the structures seen in figures 11.1 and 11.2, but this also went unrecognized.

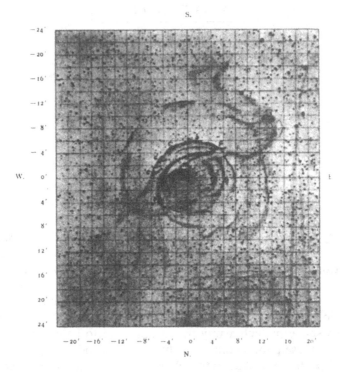

FIGURE 11.1. The light echoes from Nova Cygni 1901 as photographed on September 20 of that year. The light flash for the nova itself was no longer visible, but light reflected off the surrounding interstellar dust is visible as "ripples" reaching into space. Compare with Figure 11.2. (From the *Astrophysical Journal* with thanks to James Felten)

More than a decade later, when the distance to the nova had been derived in another way, it was realized that the speed of the apparent expansion in the nebula was indeed greater than the speed of light. The geometry of the dust requires a sheet to lie somewhere close to and in front of the nova. As light travels outward, it first illuminates a section of the dust. The light from that segment of dust then starts on its journey to earth. However, the light pulse from the nova continues to move outward and then illuminates another segment of the dust sheet. Now two light beams are headed toward the earth. They started their journey at very nearly the same time, but seen from earth they appear to come from well separated points in space. If this geometry is not recognized, one concludes that the nebula is expanding faster than the speed of light. In the case of Nova Persei the expansion appeared to be four to six times faster, an impossibility. This illusion of superluminal velocity was pointed out by P. Couderc in 1939 and only recently has James Felten of NASA Goddard Spaceflight Center again drawn attention to this work. The dis-

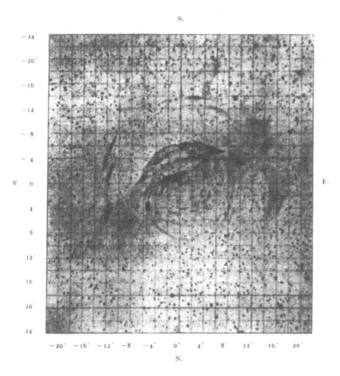

FIGURE 11.2. The light echoes from Nova Cygni 1901 as photographed on November 13 of that year. In 1901 astronomers were at a loss to account for the rapid changes in the structure of these light patterns. It would be three decades before the idea of light echoes would be accepted, only to be forgotten again. In addition, knowledge of the existence of interstellar dust, which reflects the flash of the nova, was nonexistent at the time. (From the *Astrophysical Journal* with thanks to James Felten)

covery of superluminal motion associated with Nova Persei had been all but forgotten; to the point that radio astronomers who observed superluminal motion in quasars in the 1970s thought theirs was the first discovery of the effect, as is widely reported in texts and articles on astronomy.

The little oddity associated with Nova Persei thus turned out to be a galactic light show produced when light from the nova reflects off surrounding interstellar dust to momentarily light it up.

Echoes of Light from the Milky Way

An extension to vast dimensions of this unusual reflection of light from dust clouds has been observed by Allan Sandage of the Hale Observatories in California. He was undertaking a new survey of the galactic polar regions with a Schmidt camera capable of taking wide-angle views of the heavens[3] when, in the process of searching for new galaxies, he noticed faint filamentary nebulosity in his photographs. Its existence had been noticed earlier by Beverly Lynds at Kitt Peak when she studied the Palomar Sky Survey plates.[4]

One of Sandage's images is shown in figure 11.3. He was able to show that what he had photographed was reflection of light from dust well above the Milky way. The light originates in the entire Milky Way galaxy that lies spread beneath these dust clouds. This effect is much as one experiences in a city when looking up in the night when there is a low overcast. The clouds are lit with a dull glow which is reflected light from countless street and house lights.

The filamentary nature of the structures seen in figure 11.3 is taken as strong evidence that magnetic fields must cause the dust to be constrained to moving along paths defined by the field structures. Also, these particular filaments lie in a region of sky where interstellar neutral hydrogen (chapters 16 and 17) has a very similar structure. (However, it appears that in the area of figure 11.3 the dust and the HI gas actually lie adjacent to each other, quite different from what one expects according to current beliefs concerning the association of various forms of interstellar matter.)

Light Echoes from a Supernova

Echoes, or reflections of light, from dust have also been seen from a more dramatic stellar explosion, that of a supernova. The most famous of recent supernovae is the one that occurred in the Large Magellanic Cloud in February 1987. In this case an echo in the infrared was seen in late 1987 and it was predicted that a light echo should be observable as a ring around the supernova which might be visible for decades. It should be bright enough to be seen with small telescopes starting in 1988. These echoes are also due to reflections of light from the supernova explosion bouncing off interstellar dust near the dying star.

The Cool Glow of Interstellar Cirrus

A modern view of dust which has added a valuable new perspective to the nature of interstellar matter comes from IRAS, the Infrared Astronomical Satellite, a glorious example of international success in a space mission. Launched in January 1983, it functioned for ten months and produced a mother lode of data that is still being mined. The satellite was designed and built by a team of American, Dutch, and British engineers, and scientists from these nations have made the results widely available to all nations. By 1987 is was difficult to find an astronomer

FIGURE 11.3. A high-contrast print of the filamentary and very diffuse dust illuminated by light from the Milky Way produced by Alan Sandage. These dust
filaments are found to be anti-correlated with interstellar HI filaments in this
region. (Alan Sandage and the *Astrophysical Journal*)

who was not relating his or her results to those obtained from IRAS,
whatever the field he or she was working in.

Because infrared radiation (heat) does not readily penetrate the earth's
atmosphere, astronomers have to reach into space to observe this radia-

tion. IRAS did this with stunning success. Its survey of the entire sky revealed half a million discrete sources of infrared radiation, including protostars (chapter 21), galaxies, and starburst galaxies in which enormous numbers of stars suddenly appear to have been created. It also discovered the existence of the glow from warm (20 K) interstellar dust over much of the heavens.

Figures 11.4 shows an IRAS image obtained at a wavelength of 100 microns (μm). The nebulous structure revealed in this infrared image is

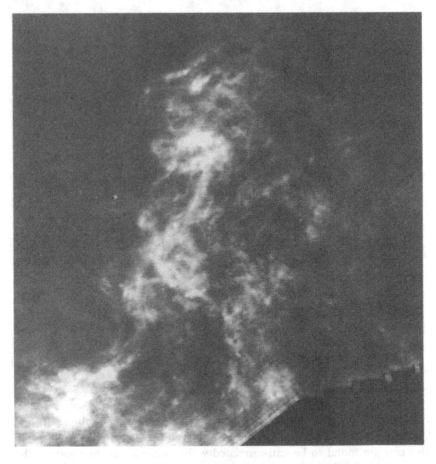

FIGURE 11.4. Interstellar "cirrus" as observed by the Infrared Astronomical Satellite (IRAS). This image made at a wavelength of 100 microns reveals emission from coll interstellar dust at about 20 to 30K. The wispy nature of this matter gave rise to the term *cirrus*. These are diffuse clouds of interstellar dust not readily observable by optical telescopes. IRAS found that the cirrus is very common close to the Milky Way. (IRAS Plate 1, centered at the north celestial pole, declination +90°, size: 15° on a side. Infrared Processing and Analysis Center, Caltech)

due to the existence of cool, very diffuse interstellar dust at a temperature of about 20 K. This discovery of this material over a large fraction of the sky caused a new name to be coined—infrared cirrus. Unlike Barnard, who was seeing the dust as dark "voids" between the stars, IRAS sees the dust as glowing objects, the reason being that matter at 20 K emits radiation which peaks in brightness in the infrared part of the electromagnetic spectrum.

The infrared (IR) emission from the relatively local cirrus reveals structure which are otherwise invisible; this dust is far more diffuse than in Barnard's dark clouds. It seems to exist nearly everywhere. It reminds us of Barnard's suspicion that there was a background glow of very faint light everywhere, not unlike the glow of the cirrus at infrared wavelengths. The cirrus is too diffuse to produce a notable effect on starlight and whether it may faintly reflect some the light remains to be seen. At present the IRAS images are the only way to recognize its existence. The cirrus clouds have already been associated with various interstellar gas clouds (to be discussed later).

Further examples of IRAS images of infrared cirrus will be scattered throughout the subsequent chapters. This cirrus is seen as luminous material over much of the sky which contrasts with Barnard's discovery that the dust absorbs light from many directions. Yet the material that is the cause of both phenomena is the same interstellar dust, luminous to infrared telescopes and dark to the optical.

Notes

[1] These images were obtained by G.W. Ritchey in 1902 and were published in *Astrophys. J.*, 15:128, 1902. I am indebted to James Felten for bringing these images and the phenomenon of the nebulosity around Nova Persei to my attention.

[2] Hinks, A.R. (1902) "The movements in the nebula surrounding Nova Persei." *Astrophys. J.*, 16:198.

[3] Sandage, A. (1976) "High-latitude reflection nebulosities illuminated by the galactic plane." *Astron. J.*, 81:954.

[4] The Palomar Sky Survey is a complete set of photographic plates which covers the entire sky visible from Palomar Observatory and is used as a basic resource and reference by astronomers. It was produced in collaboration with the National Geographic Society and is available in most major astronomical libraries.

Part 3   Interlude

The genesis of the idea that interstellar matter exists was marked by many moments of discovery, each of which brought pleasure to the astronomer concerned. But research does not always lead to happy endings.

Barnard's dilemma was archetypal. Pioneers in science operate in a world of concepts and thus have to confront previously held ideas (beliefs) about the way the universe is. They must then dare to suggest it is otherwise. The process whereby scientists do so highlights the great difference between the scientific endeavor and other human disciplines where insight into the nature of things is sought. Proof is essential in science and therefore the astronomer has to prove to his colleagues that a discovery is meaningful and in so doing must show others how they, too, may arrive at the same conclusions. Proof, however, usually depends on having data or being able to demonstrate in a logical manner how one can move from one precept to another. Intuition may create the platform from which the scientist originally leapt, but is not enough to prove anything.

This highlights the difference between the way a scientist is expected to deal with insight as opposed to the religious seer or mystic, which, in turn, highlights the gulf between science and religion. These are a few of the topics to be touched upon in this interlude.

Then we will explore further and see how, in order to bring the enormity of space and time, and the scale of astronomical phenomena, under some degree of mental control, astronomers make models. These models, in turn, almost always involve the equating of forces within a system that is in a state of balance or leads to a state of flux. Details may vary, but the conceptual steps can be understood without the reader having to examine all the details. Then consider the rest of the book akin to wandering into an apple orchard. You can pick an apple here and there and enjoy its taste without having to know all about farming, pruning, fertilization, and the general care and feeding of trees.

CHAPTER 12

The Thrill of Discovery

The scientist does not study nature because it is useful; he studies it because he delights in it, and he delights in it because it is beautiful. If nature were not beautiful, it would not be worth knowing, life would not be worth living.—Poincaré.

Eureka!

Astronomers delight in studying distant nature, the contents of the starry universe. But when you ask individuals why, you will find that their greatest delight comes from discovery. Finding or understanding something for the first time, something no one else has ever seen, provides the real reward. This delight is personal and only partly related to whether nature is also sensed as beautiful.

Scientists love to "boldly go where no man has gone before."[1] The act of discovery produces a heady sensation akin to the thrill of victory for an athlete. The thrill may come as a surprise, as the result of a serendipitous turn of events, or may be experienced at the conclusion of a dedicated piece of research whose end point was always understood. In either case the joy is associated with the revelation of what was previously unknown or not understood.

Myth would have us picture Archimedes in his bathtub, solving the problem of why objects float. He rushed into the streets, naked, shouting, "Eureka! I have found it!" Witnesses concluded that he had lost it, but the moral of the story is clear. This was his moment of joy that accompanied insight, a moment associated with the thrill of discovery.

Abraham Maslow, the psychologist, has described a *peak experience* as "the single most joyous, happiest, most blissful moment of your whole life."[2] This is precisely how the astronomers I interviewed described the thrill of discovery. Comparison of the two phenomena suggests they are indistinguishable.

Consider why you believe astronomers are involved in research. Reasons you may have heard always have an idealistic air about them. "To learn about the secrets of creation." "To aid in mankind's search for knowledge." "To understand the origin of galaxies." And so on. But don't believe these types of generalizations. Astronomers are involved because they are addicted to the thrill of discovery. Probe beneath the surface and you find that they all have felt it at some time in their lives, perhaps even as children, and forever yearn for more. Yet they must pretend that they are dispassionate searchers after the truth. The truth is that their work turns them on!.

Maslow described peak experiences in the context of esoteric pursuits where a deep insight, a "discovery," may sometimes even lead to the founding of a new religion. The scientist who has a profound insight cannot run outside, proclaim to all and sundry that he has "seen the light," and then expect his colleagues to "believe." He has to calm down and dispassionately report what he has just understood. It is his duty to communicate the essence of his discovery, and to show others that it was meaningful, and adds to the fund of shared scientific knowledge. Yet the scientific peak experience may sometimes lead to the founding of the equivalent of a new religion. A new discipline such as quantum physics or relativity may emerge.

It was through a series of interviews with astronomers that I discovered what is here reported. Each one of them, when questioned closely, confessed to having had a peak experience which was to him or her as exciting as winning a World Series is to a baseball player. At the same time, few seemed aware how pervasive is this experience among their colleagues.

Maslow found that some people did not readily talk about their moments of greatest joy or ecstasy. "Many were embarrassed or ashamed because these experiences were not 'scientific'."[3] This is also true of certain astronomers, but once they realized that others had the same thing happen to them they became open and forthcoming. Furtively some of them whispered, lest the confession be overheard, "The experience is better than drugs or sex." Or, "I don't know about drugs, but for me there has been nothing like it in all my life." Another pointed out, regarding the intensity of the feeling, "If it happened too often I'd burn out in a few months."

The thrill of discovery can be experienced in a different context, as an astronomer/teacher once reminded me. "The first time it happened to me, even before is happened in my research career, was when I had to teach. Every time I understood another piece of what I had to teach I felt elated."

The thrill of discovery or of experiencing insight or understanding can propel someone into a state of ecstasy, euphoria, or elation which may last minutes, weeks, or months. The memory lasts forever. Once someone has tasted the thrill of discovery he/she will devote the rest of his/her life to the pursuit of more of the same, and be willing to do so with amazing dedication. Is this not a benign form of addiction?.

A heartening fact, for those who have not yet had the thrill, is that experience does not depend on having lots of qualifications, training, or even brilliance. It is given as a reward for curiosity.

A Few Examples

Edward E. Barnard was twenty-three years old, uneducated and struggling with poverty, when he experienced his moment of transformation. With the assistance of J. W. Braid, instrument maker and colleague at the photo gallery where Barnard worked, he built a small telescope which he turned on the moon, Jupiter, and the double star in Ursa Major. He was thrilled by what he saw.

"This simple telescope," reported Braid later, "gave Barnard more pleasure than *anything else in his whole life*" [my italics]. The hyperbole makes the point and may well have been true. Barnard went on to make dozens of original discoveries, each of which must have been a tremendous thrill for him. He discovered seventeen new comets, the fifth moon of Jupiter (Amalthea) and hundreds of dark markings among the stars known to be interstellar dust clouds. Why did he spend each clear night for the rest of his life looking through a telescope? Because he was addicted to the thrill of discovery.

Probe beneath the surface of the researcher and you will find that many, if not all, astronomers have had experiences similar to Barnard's. Maslow found that those who did not readily admit this were the same individuals who had difficulty in expressing feelings. They unconsciously pushed away memories of their most exciting moments. He also discovered that at some time or other most of us have had a profound peak experience, a transcendent moment, which can change the course of our lives.

An astronomer recalls a night when he was twelve years old. He was looking at Saturn through a small telescope. The night was partly cloudy and the planet appeared but briefly. As the child gazed upon the magnificent rings he suddenly felt that he was the only person in the world who, at that very instant, was looking at Saturn. It matters not whether he really was; what matters is that he believed it. He described the sensation. "I suddenly felt one with the earth and the universe. There was nothing else." A wave of elation and euphoria swept over him and decided his career. This was the peak experience in full cry. "I knew then that I would become an astronomer, no matter what it took."

Maslow reported that peak experiences have the characteristics traditionally associated with moments of deep religious insight, as culled from the recorded history of many creeds and faiths.[4] William James described this as a "mystic experience", Freud called it "the oceanic experience" and Richard Bucke labelled it "cosmic consciousness."[5] For the scientist it is the "eureka experience." Individual descriptions may differ in detail, no doubt determined by the leanings of the reporters, but they are all manifestations of the same phenomenon.

The great difference between the scientific endeavor and the esoteric tradition is what the experiencer does *after* having the sensation. For the esoteric and his followers it is enough for him to say that he has "seen the light." He does not have to prove anything to anyone. His followers need only believe. But the scientist must explain what the light has revealed. Otherwise the experience has no significance beyond the personal thrill induced. The scientist must find the words to enable him/her to report what it is he/she understood or discovered so that others may achieve an equivalent level of understanding. Thus science as a discipline moves forward, evolving through sharing, while religions and sects remains static, rooted in the words of a founder who had to "prove" nothing to anyone.

An astronomer recently spent years analyzing all data related to the structure of the spiral arms in the Milky Way. Day after day he worked at it. Then suddenly it happened! "One day I heard the Galaxy," he told me excitedly. "I hesitate to admit this, but I heard the music." He stared at me, wondering if I would think him crazy. "I could hear the music of the spiral arms. They have motion within them and I could hear it! It was an incredible feeling. I understood!" He confessed that it was the greatest moment of his life, could not tell his colleagues, and wasn't optimistic that anything as wonderful would ever happen to him again.

Then he had to confront the astronomer's next great challenge. After the elation wore off he had to write a report on his research. Today his neatly rational paper on the structure and motion of the spiral arms rests between the covers of a journal in astronomical libraries throughout the world. The report makes no mention of his moment of profound insight, nor of the music.[6]

Frank Goble noted that "the peak experience may be brought on by many causes: listening to great music, a great athletic achievement, a good sexual experience, even dancing."[7] And, we must add with considerable emphasis, by the thrill of scientific discovery, by the satisfaction of curiosity that accompanies understanding, by the completion of a piece of research. Seeking this feeling may be the strongest driving force behind the scientific endeavor, but it is never mentioned, because, as Maslow put it, the experience is intrinsically "unscientific."

Recently astronomers at the Naval Research Laboratory in Washington, D.C., reported the discovery of a fascinating new phenomenon (see Chapter 22). A dark, unseen, mass had moved in front of a distant quasar and interfered with radio waves on their way to earth. A few months later everything was back to normal. The moment of recognition of this phenomenon (called an *extreme scattering event*) is etched vividly in the mind of Ralph Fiedler. It was a moment that culminated months of hard work for him. "I was elated and ran screaming and shouting down the corridor," he says, expressing a feeling to which many a researcher can relate. But his elation had to be tempered so that he and his colleagues could take up the challenge of explaining what had been observed. And when the explanation was found the thrill returned; a little less intense perhaps, but still a delight. (The dark masses responsible for the phenomenon turned out to be diffuse clouds of hot gas having planetary dimensions which are more numerous than stars in the galaxy.)

Most amateur astronomers have experienced the thrill of discovery when they spy something they have never seen before, or when the sky allows a particularly good view of a familiar object. Why else would they peer through telescopes on cold nights or gather in remote places to share talk and look at stars? Because, to paraphrase the words of Poincaré, they delight in it. It feels good! That is why amateur astronomers love the skies and professionals explore the unknown. The moments of success, the moments when you see something for the first time, feel very, very good.

The discovery of a supernova, the explosive death of a star, has provided a rare thrill for only a handful of human beings in recorded history. Not since Tycho Brahe in 1572 and Johannes Kepler in 1604 had anyone actually seen a supernova; not until February 24, 1987, that is. On that night, at Las Campanas Observatory in Chile, astronomer Ian Shelton discovered a star in the act of exploding. Unlike Brahe and Kepler, who did not know what they were witness to, Shelton knew what was he seeing.

"I don't think anything is going to replace that night of actually seeing it for the first time. That was memorable." These were Shelton's words on the *Nova* television program describing the discovery of supernova 1987a. He said he had gone outside to look at the Large Magellanic Cloud because a new star had appeared on his latest photographic plate of that galaxy. He wanted to make sure the new object was really there, and it was! This was a moment of discovery countless astronomers have dreamed about.

As the narrator summarized; "Ian Shelton's place in history is secure but perhaps that is less important than the thrill of discovery itself."

The thrills provided by 1987a were not solely Shelton's. Later, others independently discovered the object and the thrill was theirs as well, even if their names will not be recorded for posterity.

Not least of those who were elated by the event was astrophysicist Stirling A. Colgate of Los Alamos. For decades he had been pushing a theory that supernova explosions required the generation of neutrinos. When he heard that these elusive particles from the explosion had been detected at the neutrino observatory at the bottom of a zinc mine in Japan, he had his peak experience.

"It was far out," Colgate exulted on camera. "Then this theory of so many years, of a lifetime, was suddenly out there (manifested in the explosion seen by Shelton). What a trip man! What a real trip!" He bounced around in excitement. "It was just ... you know." He shook his head in frustration because words failed to convey the thrill. "Oh, I can't tell you any more." With that he walked off camera, a beautiful illustration of the thrill of discovery and what it means to the individual.

The thrill of discovery for the astronomer is as wondrous as the thrill of victory for a sports star. Following a Super Bowl we are treated to TV images of bottles of champagne publically shaken and jets of bubbly squirted over victorious players milling about the locker room. An interviewer struggles, amid a rain of alcohol, to get a coherent word from the MVP.

"This is the greatest day in my life!" the linebacker burbles excitedly. "It is a moment I never dreamt would happen," the tight-end sputters, eyes glazed, senses reeling, "I feel great!"

The hapless viewer may be forgiven for thinking that such thrills are beyond the experience of mere mortals. But for the professional astronomer who has just made a discovery, or the amateur who sees a distant galaxy for the first time through his homemade telescope, the feeling is every bit as wonderful. They may also break out the champagne, but have more sense than to pour it over someone else's head.

Just Say Yes to Curiosity

The thrills and the rewards that accompany scientific research are personal and scientists have conspired mightily to keep this a secret. They have come to be pictured as slightly inhuman beings in white coats pursuing lofty goals for reasons no one understands. Yet they are human, and the truth is that scientists "get off" on what they do.

Everyone can experience the thrill of discovery, even if you don't have a Ph.D. All you need do is exercise your curiosity and be willing to recognize that there are things in nature worth discovering for yourself. So why not say yes to curiosity? Embrace those moments when questions presents themselves. Seek solutions and find out how beautifully elegant the world of nature is. Prepare to enjoy the thrill, the "high," that is, the reward. It could present you with the "single most joyous, happiest, most blissful" moment of your life.

Notes

[1] Star Trek.

[2] Maslow, A.H. (1962) "Music education and peak experiences." *J. Humanistic Psychology*. Vol. 2.

[3] Goble, F.G. (1962) *The third force*, New York: Grossman Publishers.

[4] Ibid.

[5] Bucke, R.M. (1977) *Cosmic consciousness*, Secaucus, New Jersey: Citadel Press (Reprint of his 1900 book).

[6] The reader will appreciate that most of these "confessions" were obtained with the promise not to reveal names, a sobering comment on our shared attitudes toward the conduct of scientific research and the image that researchers find they need to uphold even in the eyes of their peers. This topic surely deserves further, more formal, research.

[7] Goble, op. cit.

CHAPTER 13

Science and Religion

Dealing with the Experience

There exist deep similarities between the reaction of the scientist to his peak experience and the mystic to his religious insight. They both feel an illumination, an awakening, and in both cases in regard to very specific and, one might add, narrow aspects of life. The religious seer will want to rush down the aisle of the church and shout out, "I have seen God," because other words fail to convey anything of meaning. The scientist following a successful detection of a new phenomenon for the first time in human history may have the same urge, to run "shouting and screaming down the corridors" as it were. But he is aware that it cannot end there. There are well trodden paths down which the scientist must walk in order to communicate insight and discovery to the rest of the scientific community and then to the world.

The religious peak experience, as Maslow has pointed out, may form the foundation for a new religion, but in the end the experience itself is lost. Organized Religion, the greatest enemy of the religious experience, then takes over. Richard Bucke, in *Cosmic Consciousness*[1], relates stories of nearly four dozen people who experienced profound cosmic insight, from Jesus to Walt Whitman and including many lesser-known individuals. The lesson of his stories is that we can all have these experiences. The irony is that organized religion has for a long time regarded as heresy the notion that you and I might be capable of having cosmic insight, a "Christ-like" experience, as it were.

Organized science, on the other hand, has created a system in which we are free to search for knowledge while the personal peak experience associated with uncovering such knowledge is underplayed, if not entirely ignored, in public discussion. When the researcher writes the report he must leave out subjective descriptions and reference to feelings. The reasons are simple. Scientists have a greater chance of getting at the underlying truth about reality if they rise above subjectivity; otherwise their wishes, hopes, and expectations blur their vision, even to the point where they cannot see.

The corollary is that the experience of insight, the moment of understanding or discovery, the profoundly personal, exciting, and elevating instant, must be put aside because it contains no information about the nature of the external world. The scientist has to calm down and make sure that his report of the experience has some relationship to reality (here defined as the universe external to the imagination).

When human beings only heed the subjective content of the peak experience they may end up chasing the insane dreams of another. Witness the tragedy of the Jonestown massacre, when 400 people committed suicide because they "followed" the vision, the "insight," perhaps, of a deluded man.

Herein lies the profound demarcation between science and religion. The scientist must set aside the personal reaction experienced in a moment of insight and strive to reveal the truth behind it. The mystic, however, is encouraged to focus on the personal and need not search for objective content beyond the experience itself.

Organized science has evolved to the point where its "rituals" go beyond the subjective peak experience, even if such experiences were the root of the inspiration or insight and a demonstrable breakthrough in knowledge. Organized science recognizes that the subjective awareness of the universe and its contents lie in what you *want* to believe, rather than being coupled to what *is*. Thus it cautions its practitioners to beware. Many of them have come to beware too much and fear the wellspring of their own creativity.

The success of the scientific method has been to reveal that there really is something beyond subjective experience, that there does exist a physical universe worthy of study. This is contrary to what certain mystical traditions, such as Buddhism, teach. They question the very existence of a universe beyond the senses and intellect, implying that the only universe that exists is the one we create. This view of nature reflects no more than primitive intellectual laziness.

We must note that the use mankind has made of scientific knowledge about reality is not always of the highest moral or ethical caliber. This is not a fault of science, per se, but of fallible human nature. After all, human beings have only just emerged from the cosmic cradle, and who said they were perfect?

Rituals

The rituals of science act to dampen individual ardor when it comes to reporting truth. To do otherwise would be to welcome snake oil salesmen into the laboratory. If we ceased to take a critical view of the contents of someone's "insights," imagine the chaos that would result. One conjures up visions of countless sects within astronomy. Believers in Herschel's opinion that there were holes in the heavens would do battle in the observatories with followers of Ranyard and Curtis who believe that dark matter exists between the stars. In this scenario neither group would further study of the phenomenon because they would be too locked in dogma, and hence in combat.

Another comparison between science and religion reveals that the scientist experiences a great freedom forbidden the religious seeker. The former is encouraged to study the consequence and meaning of someone else's peak experience and is expected to duplicate the insights and awareness for himself before moving into unexplored territory as a researcher. A similar behavior is not encouraged amongst fundamentalists. To test whether what Christ taught makes sense is obviously not encouraged. We draw this conclusion by simply looking around us. To suggest we might explore further than Christ did, or even duplicate his findings, is to risk being locked in a "safe" place.

By way of illustration of this point, consider that while Barnard was the first to see the dark clouds, and that other astronomers proved that they were interstellar matter, the modern student of the subject is expected not only to test the truth of this, but to go beyond what is known; that is, to do research and explore further. The student of science should not stop after studying the words of the masters. He or she is encouraged to push back the frontiers of knowledge and if he or she happens to prove the masters wrong, so be it. If scientists were suddenly to adopt the mechanics of the religious syndrome, someone such as Einstein would be immediately elevated to sainthood and no one would be allowed to question his work or encouraged to do research into the nature of relativity or gravity.

Organized religion insists that adherents *follow*, which, by definition, means the follower cannot experience what the founders, the masters, did. This tragic limitation, a phenomenon in all major religions, must inevitably spell the extinction of the organized church, because human beings are innately curious and sooner or later their curiosity will carry them into new realms of insight and understanding. The apples hanging from the tree of knowledge inevitably lure us to greater awareness of our relationship with the rest of the universe, invariably at the expense of old notions as to the nature of reality.

The creative student of science is not expected to *follow* anyone beyond a certain point The study of interstellar matter, for example, would still be in the nineteenth century if astronomers had not looked beyond Herschel's words or what Barnard discovered. The astronomy student must learn what the great figures did, but if they choose research as their career they cannot stop there. The goal is to move ahead and go where no one has gone before. Nobel prizes are symbolic rewards for those most successful in this adventure.

Religions, therefore, are constituted by legions of followers who dare move no further than their founders. The legions of science, on the other hand, are expected to be doers whose primary goal is to see further than their predecessors; by standing (to paraphrase Newton) on the shoulders of giants, if necessary.

Addiction

The scientist and the mystic both have peak experiences that drive them. Maslow has argued that the mystic may be tempted to seek further peak experiences and "... to value them exclusively, as the only or at least the highest goods of life, giving up other criteria of right or wrong. Focussed on these wonderful subjective experiences, he may run the danger of turning away from the world and from other people in his search for triggers to peak experiences, *any* triggers."[2] The scientist is often also guilty of turning away from the rest of the world in his lonely pursuit of the experience. The mystic, according to Maslow, may lust after more of the experience as an addict seeks his highs because the peak experience related to the satisfaction of curiosity brings about a state not unlike a drug high.

If one is impatient one can turn to drugs for a quick repetition, but this produces an artificial result. The research scientist, on the other hand, is quite capable of laboring for decades in search of apparently and relatively insignificant goals from whose realization he will receive enormous personal satisfaction. Even a simple breakthrough, as seen from an outsider's point of view, may create in the patient researcher a sense of ecstasy, even if it is in the quiet satisfaction of a job well done.

Such psychological rewards are profoundly satisfying even if the only value attached to them are set by the individual. For example, one person's exciting new variable star observation may be another's tediously obtained, perhaps boring, piece of information. Another's ecstatic measurement of the temperature of the universe may, to another, appear no more than a punctuation mark in the course of astronomical history.

The scientist reacts in a dual mode to the peak-experience. What he feels upon discovery is pleasurable, but also alien to the idealized expectation of what it like to be a scientist. This reinforces his reaction against subjectivity, because he has now been guilty of that very sin to which, he was taught, as a youngster perhaps, that good scientists didn't succumb. His own thrill of discovery produced a tremendous sense of personal and extremely subjective pleasure. The experience may even have come as a terrible shock. It might even have been frightening had he heard the music of the spiral arms and not been prepared to handle it. As a reaction, some scientists may develop an even greater "rationality" which then protects them from further cases of the very experience that hooked them into their careers in the first place.

Yet the quest continues. Once the peak experience is tasted, the desire to have it again subtly diffuses one's being. Maslow writes that in due course the religious seeker requires stronger and stronger stimuli to produce the same responses. Here we recognize the drug addict's behavior as he grows resistant to the chemical. More and more is required to produce the same "high." Since chemical drugs are harmful to the physical well-being of the individual (and society), society cannot condone this path to the peak experience, the moment of insight. However, the religious quest is so focussed on the experience itself that it drives the individual ". . . into the magic, into the secret and esoteric, into the exotic, the occult, the dramatic and effortful, the dangerous, the cultish" in pursuit of stronger stimuli.

A parallel might exist in Big Science where teams of physicists get together to build a $6 billion dollar super collider so that they can experience the ultimate thrill—understand, for the first time, the most fundamental secret of universe. What does matter consist of? Can there be a more seductive lure for any scientist, to be the one that comes up with the answer? Just how much is society willing to pay for this "high"?

The Urge to Explore

We cannot underestimate the power of the human drive to explore the unknown, to search for discovery and understanding. There may be an important evolutionary reason that we experience pleasure from the satisfaction of curiosity. In fact, this drive may be at the very root of evolution, for curiosity has been found in virtually all animal species.[3]

Maslow notes that for the religious seeker, "the peak experience may then be exalted as the best or even the *only* path to knowledge, and thereby all tests of the *validity* of illumination may be tossed aside." But organized science has structured itself so that verification is essential before one's insight is accepted into the "scientific hall of fame," the repository of *true* knowledge. Any individual may have a peak-experience, but the content of the experience is not necessarily related to external truth. The scientist's lot is to go beyond and share what he experienced in words others can understand.

Inventing New Languages

Bucke and others have reported that the sensation of cosmic consciousness is usually described as being beyond words. Is that because the person doesn't know how to use words, or because the appropriate words haven't been invented yet? If it is the latter we can take heart in the fact that there were times when no human words existed, and just because the necessary words and concepts don't exist now doesn't mean they won't be suggested in years to come. We just have to keep working at it.

The scientist often has the same problem of communicating the nature of his insight to a world that might not understand. So it was when the concept of gravity struck Newton. Unlike the mystic, he did not slump under the apple tree with a lump on his head and say, "I understand now but I can't put it into words." He didn't shrug apologetically and mutter, "I understand why the moon is in orbit and why the apple fell to earth, but I can't tell you why." He proceeded to invent a new language that allowed him to communicate about his discovery and to explain the force of gravity to those who wanted to understand. The language was called *calculus*.

Mathematics is a language. So, each in its way, is physics or chemistry or biology. They are languages that help us communicate about certain aspects of the universe. No more and no less. Herein lies the strength of science. Newton described his insights into the nature of gravity with a form of mathematics that he had to develop after the apocryphal apple went bouncing off his head. Meanwhile the mystic sits in his ashram and insists to his adoring followers that words cannot describe his experience. The irony is that his followers must believe him, or they cannot be followers!

Scientists have left the mystics far behind when it comes to communicating what it is they have experienced, while the mystics still stutter that their peak experiences are beyond words. Too bad!

The astronomer who observed the world's first pulsar may have felt a profound sense of awe, but organized science encouraged her to communicate the discovery. She wasn't limited to using the English language. The vocabularies of physics, mathematics, chemistry, and the other sciences were available for the task. Rest assured that when spoken correctly, others who have learned those languages will get the point. They, too, will "see the light" and understand that which was originally beyond words.

Barnard's struggle with the meaning of the dark markings among the stars was due in part to the language which enabled astronomers to clearly recognize the existence of interstellar dust having not yet been developed. Astrophysics was in its infancy, and it would take more than just photographs to prove that there was matter between the stars. The language, the concepts, and the appropriate descriptions had to be invented, one step at a time.

Today we *know* there are dust particles between the stars. Further research may refine what we know about the particles, but such research will never erase the concept of interstellar matter from our awareness. Research will refine and redefine what we know about the details. The subject will evolve and it is the scientist's duty to see to this evolution.

And so, in this book, we will soon explore new aspects of the language of science in order to communicate what is known about other aspects of interstellar matter. New words will be introduced and each will refer to something—a phenomenon, process, or class of objects—not known a half-century ago. We will meet interstellar molecules and spectral lines and protostars and giant molecular clouds. When you begin to *feel* what these things mean in the context of the new language, a whole new universe will open to your imagination and understanding. You will see things you could not comprehend if you depended only on English or Russian or Swahili. Scientists have known this for centuries. They do not hesitate to invent new words, concepts, definitions, languages, or even dialects, in order to communicate about what they have discovered. In the pursuit of scientific knowledge and understanding nothing is sacrosanct, not even language; certainly not dogma.

The Eternal Quest

A tragedy of our age is that a huge segment of our population never has a chance to discover what is going inside the corridors of the observatories and the laboratories where the evolution is taking place. Thus the disillusioned turn to the Maharishis and the Rimpoches and rebel against Science, implying that Organized Science is out of touch. Yet the great scientists have always known that they are driven by a metaphor of the religious urge. However, their religion (science) has never become immobilized through adherence to archaic symbolism that serves only to hold back the evolution of ideas, and hence knowledge. Such evolution is a direct threat to dogma, for evolution implies change, which must ultimately spell the end of current dogma, whatever its form of expression may be.

The voices of insight experienced by the mystic cannot be confirmed or tested, but this does not invalidate them for that individual, only as a way to obtain useful information beyond that individual's mind.

The concept of good and evil, and how to distinguish one from the other, now also raises its head. What is right and what is wrong? Organized science has taken care of this matter for its adherents by instituting simple tests. Are your observations confirmed? Can a colleague duplicate your experience, that is, your experiment. Can she repeat the measurements and obtain the same results? Can others? If so, we are all closer to the truth.

Is it possible for another theoretician to follow her way through the maze of your equations and see for herself the truth and beauty you claim to have found in your latest version of the unified field theory? If not, then try to find where you went wrong and try again.

Yet the mystic and the scientist are engaged in the same quest. They initially walk the road together until they come to a fork marked by a signpost. The mystic stops, sits cross-legged and begins to meditate upon the meaning of the signpost. The scientist proceeds along the road to explore where it leads. If he should arrive at a dead end he will return, choose another path, and explore that.

To become enlightened you do not have to sit cross-legged like Buddha. You are allowed to move about. Everyone is allowed to question the universe, exercise his curiosity, and then examine critically what the quest reveals. If you are lucky enough to make an astronomical discovery in the process, why not lay a copy of the *Astrophysical Journal* at the Buddha's feet and wish him well? Then proceed once more along your path.

I feel deeply frustrated when observing the dichotomies that exist between organized science and organized religion. However, I am heartened by the fact that scientists have made demonstrable progress in understanding the nature of the universe and of life. The scientist's work can be verified and has revealed shared truths. On the other hand, a cursory view of history immediately reveals that the dogmatists have done little more than kill each other off in the name of their beliefs, which are far from *shared* truths. Their beliefs are usually founded on someone else's peak experience reported thousands of years ago and the adherents are to be found sitting with the Buddha at the signpost (ignoring the piles of scientific journals spread before them!).

To return to Maslow, "small r religion is quite compatible, at the higher levels of personal development, with rationality, with science, with social passion."[4] In other words, there are no "two cultures" when we deal with the experience of being a scientist or a mystic. The dichotomy enters when we compare organized science and organized religion. The former is a newcomer on the intellectual scene and the latter is threatened by the youngster's power, a power that is rooted in its demand that a personal experience must, and should, evolve to become a shared truth if it is to have meaning. Science insists that we show others how to have the same experience, to share our insights, and to find the language to do so, creating one if necessary, and then to guarantee others the same results in their pursuit of truth.

Notes

[1] Bucke, R.M. (1977) *Cosmic consciousness* New Jersey: Citadel Press.
[2] Maslow, A.H. (1970). *Religions, values, and peak experiences* New York: Viking Press. p. viii.
[3] See, for example, Griffin, D.R. (1984). *Animal thinking* Cambridge: Harvard University Press.
[4] Maslow, op. cit.

CHAPTER 14

Modeling

Pictures in the Mind

The reaches of interstellar space are quite beyond our keenest vision. To picture interstellar matter you have to use your imagination. That is what the astronomer also does. He will make a use of a *model*, a scaled-down version of the real thing, and then look at it and manipulate it in his mind. A model may also be a manageable version of the process one is trying to understand, manageable in the sense that it can fitted it into a mental pocket, or represented on a piece of paper, or in a computer, or on the pages of a book such as this.

Imagine a football game in progress. You are listening to a radio commentary of a game between the Denver Broncos and the Dallas Cowboys. If you know something about these teams you will quickly form some image in your head. You might picture the orange shirts of the Denver team, or see the players in your mind's eye if you know their numbers and what they look like. You can hear the crowd sounds and the commentator's voice. If he is good, and you have some idea what he is talking about, you will be able to make a model of the game in your imagination. You'll picture the quarter-back faking a handoff while the wide receiver runs downfield to take the pass. You exult (or sink into your chair and groan in despair, depending on whom you support) when the word *touchdown!* is called out.

For the reader who knows nothing about American football, this paragraph will not have conjured up significant images. He will not be able to make even an elementary model of what a football game is all about.

We make models like this in our heads all the time. It is a far simpler process to make a model of the field of play and imagine the teams moving from right to left, than to be on the sidelines for every game. For most of us it is more practical and economical to make the model in our heads and enjoy the game from there than to fight the traffic and the crowds in order to get to the field. Television, of course, makes seeing the game easier because it saves your having to indulge in any of the mental work required to make a model.

Consider another model that is part of everyday experience. You want to visit the Grand Canyon, but it is 1,000 miles away. You obtain a brochure and look at that. You might consult a road map on how to get to the canyon. You plan the journey and where you will stay when you get there, where to stop on the way for the night, which sights to take in. Inevitably you will form a mental image, a model of where you are headed and how you will get there. All models are manageable versions of the real thing which allow your mind to grasp a reality which lies beyond its immediate perception, beyond even the photographs and the maps in front of you.

Astronomers see the interstellar medium and its contents in their imaginations. They take this so much for granted that they forget to tell anyone else that this is what they do.

Like a chalkboard diagram of a football play, the astronomer's charts and diagrams express what is going on in space. But only if one already knows something about the game will the diagrams make sense. Thus you have to learn some of the basic steps, some of the rules, and then the pictures begin to have meaning and you can make your own model, perhaps seen projected on your personal mental screen or space.

Interstellar Weather

Imagine now some of the objects and events that occur in interstellar space. Plan to make your model flexible enough so that you can speed up time to suit your needs. You'll begin to see something quite remarkable. You will witness the changing weather in the Galaxy.

A front moves through space, bringing local stormy conditions. Winds sweep away from stars at enormous speeds, wrecking anything in their path, tearing at magnetic fields and clouds of gas and dust that exist there. Swirling eddies hurry between the stars even as light and ultraviolet radiation eat into surrounding clouds of gas, boiling away material that was once close to absolute zero, now suddenly raised to 10,000 K by all that stellar energy.

Blizzards of particles driven by supernovae heat up surrounding matter, driving it about violently. This is a cosmic storm. Particles travelling near the speed of light (cosmic rays) stir things up and keep everything continually mixed. There are no stagnant pockets of gas in interstellar space. The radiations and the winds driven from the high pressure regions around stars reach everywhere.

In some places it may appear relatively calm, for millions of years perhaps, with stellar light bathing all in a soft glow as gentle interstellar winds waft this way and that. During the calm the atoms might congregate on dust particles and coalesce into larger molecules, forming complex structures which are not unlike the building blocks of life on earth.

The weather in space may not at first sight be as changeable as it is on earth. However, everything out there only *appears* to move slowly. On the galactic scale of time and space the changes are dramatic. Weather fronts sweep vast volumes of space at near the speed of light, yet the distances covered by interstellar storms are so vast that millennia pass before we on earth can detect that something has happened.

It is no different for us on earth. You may be tearing across the United States on Interstate 81 at 70 mph, aware of the countryside whizzing past, but seen from an orbiting spacecraft you would hardly appear to be moving at all. It all lies in the perspective.

The terrestrial weather forecaster uses a model of the earth's atmosphere to predict where the cold front will move. Astronomers would like to be able to forecast the weather in space, even though they wouldn't be around to see their predictions come true. However, prediction and understanding go hand-in-hand. The forecaster would like to be able to say just where and when the next star will explode and cause local hurricane conditions, or where the next black hole will form to suck matter into a tornado that will form around it. They'd like to be able to foretell where the next storm of particles will emerge from a birthing star so that they can observe it more closely. That is what science is about—forecasting. But astronomers are hampered in making galactic-scale forecasts because the events they study take absurdly long to occur.

Rather than sit around and watch the life history of some region in interstellar space which may or may not form a star, astronomers search the heavens and find hundreds of examples of each of the many stages that a stellar mass might pass through: interstellar gas cloud, dust cloud, giant molecular cloud, swarm of stars, and dying stars erupting matter back into space. The experience is akin to having only a few minutes to figure out all the details of the human life cycle. You cannot wait for 80 years to find out, but if you analyze the population of a fair-sized town of, say, 10,000 people, you'll quickly learn about the cycle from birth to death because there will be many cases of each phase at any instant.

This is how astronomers have come to learn so much about the life cycle of stars and gas clouds—because there are so many examples of each stage available for study, Then it is only a matter of figuring out how they all fit together. This requires a model.

Trade winds also blow through our galaxy. Known as density waves, these prevailing winds push gas and dust around as they gust along vast fronts called spiral arms, triggering star birth in the process. We'll say more about these later.

How fast and and how far do the stormy winds in space travel? There is a simple factor that astronomers carry in their heads. At a speed of 1 km/s you will travel 1 pc (3 light-years) in a million years. 1 km/s is a lot faster than you or I will ever travel over the surface of the earth. It is 60 km per minute, or 3,600 km per hour. Travel that fast for a million years and you will have moved 3 light-years. The sun is speeding through space at 20 km/s with respect to surrounding stars so that a million years from now we will have moved 20 pc (60 light-years).

It also "rains" in space, but this rain is not in the form of water droplets that fall to earth. The stardrops are protostars which suddenly condense by the hundreds and even thousands out of giant gas clouds and then swarm about each other in clusters until they are dissipated by motion with respect to each other. The stars will literally evaporate into space, each of them an independent "droplet" of highly condensed gas.

If you take a look at the current models of the Milky Way you will see an occasional "thundercloud" of gas and dust particles rising out of the plane, rising and cooling until, under the inevitable tug of gravity, it falls back into the Milky Way again. If you could watch interstellar matter in a time lapse movie, letting millions of years go by in a minute, you'd see constant activity and upheaval with never a pause. The galactic weather is rough and stormy indeed.

Interstellar clouds are everywhere. They come and go as forces push and pull, heat and cool. Clouds coalesce, disintegrate, and reform. Some clouds disintegrate into shimmering clusters of stars. Always the clouds will obey the authority of interstellar magnetic fields which control motion and the physics of the gases. This is because some of that matter is ionized—electrically charged—and such particles respond to the influence of magnetic fields, obeying its rules as they move through space. The magnetic fields force all the particles to do their bidding. There is no equivalent on earth, for the gas in our atmosphere is neutral and ignores the earth's field.

So why do astronomers search interstellar space and make models of something which is so much larger than ourselves? For the same reason that once prompted Darwin to explain why he so much enjoyed exploring the world around him. Because, "I discovered, though unconsciously & insensibly, that the pleasure of observing & reasoning was a much higher one than that of skill & sport. The primeval instincts of the barbarian slowly yield to the acquired skills of the civilized man."[1] He spoke in particular with reference to why he preferred to go into nature with a notebook rather than a gun.

Listening to the Universe

Have you ever listened to shortwave radio? It's not anything like tuning into your local FM or AM station from your town or a city just beyond the horizon. When you tune into the shortwaves, you travel beyond your nation's borders. No customs checks or visas are required. You will travel over oceans, covering all of the planet, touching down in any nation you chose, simply by turning the dial slowly and homing in on the signal that interests you.

You can travel from the Andes to Australia, from New Zealand to China, to France, Germany, the Netherlands, and a dozen Spanish-speaking nations. Foreign voices jabber and if you listen closely you will learn to identify who lives where. To do this it helps to make a model in your head, one in which you may picture the globe and where the countries are to whose broadcasts you are tuned.

This is not always possible, though. As a teenager I'd occasionally listen in fascination to snippets of a commentary of an American football game, without being able to understand any of it because I had no idea of what such a game looked like. I could not even begin to make a model!

Now imagine tuning your radio set even further. You'll need a large antenna for this; not just a piece of wire strung along the wall. A large dish is necessary. Then you will begin to tune into stations way beyond our planet, beyond the solar system, out in space between the stars. You'll be able to hear the stars themselves. But none of these astronomical radio transmitters broadcasts music. There is no entertainment, per se, unless you are ready to learn what the signals mean. Once you have learned the language of radio astronomy those signals become very entertaining indeed. Some of them, like the beat of the pulsars, may even sound like avant garde music, but otherwise the cosmic radio waves have little to provide by way of directly stimulating entertainment. Only when you become actively involved in interpreting the signals will they become exciting.

The symphony of a billion stars and a million galaxies pervades the air waves. You need a large radio telescope to tune in, to listen. Such devices cost many millions of dollars and there isn't one at your local cable company premises. However, even if the National Radio Astronomy Observatory was to pipe the signals they receive into, say, channel 214, you'd quickly tire of the constant hiss you'd hear. Best leave it to the professional to interpret those hisses. Let them make the models which they make available for us all to see. That saves us all a lot of bother.

Making Temporal Models

Astronomers love to run time-dependent models. These can be very sophisticated and their essence is that with every passing second the model reveals some new aspect of the physical system being modeled. For example, models of stars simulate the physical and chemical processes involving millions of particles interacting at temperatures and densities that depend on depth in the stellar atmosphere. When the computer "runs" the model the parameters initially fed in as data change because all the equations in the computer are of the form $A = f(t)$. This means that some parameter, A, has a value that is a function of time, t.

Astronomers run a time-dependent model and can watch a star evolve and then discover that it suddenly explodes; the model star explodes, not the computer. The explosion is witnessed in the numbers displayed by the computer. Nothing melts. No alarm bells sound. No lives are lost from buffeting by the supernova shock waves. The data displayed on the screen or the printout say it all. They are enough, and make modelling a fascinating process that avoids all that sitting around and waiting for a real star to explode.

Model making is one of the most time-consuming and passionate interests of a majority of theoretical astrophysicists. Models allow them to make predictions that can be compared with reality. When the predictions are correct, understanding results—and that produces a variation on the thrill of discovery.

A Question of Balance

Astrophysical models are particularly fascinating and informative because they use equations which involve parameters that vary in time. Since equations require the expression of something on each side of the equality sign, this implies that the models deal with questions of balance. In astrophysical equations this generally concerns the equating of forces that oppose each other, those acting to disrupt versus those wanting to unite. The equations of balance describing an astronomically interesting situation or object will usually compare forces acting inward to those acting in the opposite, outward, direction. The primary force of coalescence driving matter together is gravity. In addition, external pressure acts to push inward on the object in question.

Forces acting to disrupt include the motion of particles in the cloud or star, a motion that is controlled by the temperature of the object. Rotation of a cloud or star will also produce a force which acts outward, as do the large scale random motions of "blobs" of matter within a cloud. This is called turbulence.

There exist other forces, such as are produced by the interaction of matter with magnetic fields, but as yet their role in astrophysical situations is far from understood.

Thus in the study of interstellar matter, as well as in most other branches of astronomy where sufficient data are available, the astrophysicist may explore whether the action of gravity due to a certain mass of material in a cloud is sufficiently great to overcome the disruptive effect of heat in the cloud. If so, the cloud must be contracting, something that has been inferred without having to sit around and wait for the cloud to grow smaller. Should the cloud contain more energy than can be held under control by gravity, the thing will expand and disrupt. Exploration of the amount of imbalance in the equations then tells how fast the expansion proceeds and, again, no one has to sit and wait around for the cloud to disappear because the researcher can state that it will be gone in 10 million years, for example.

The Chinese symbol for balance is that of yin and yang, opposites in harmony. Very seldom does one find perfect stability or equilibrium in interstellar matters, however. In astrophysical situations, when yin and yang are unequal, the system (cloud, galaxy, or star) will evolve. Stars, however, are often locked into eternities of stability, a fortunate circumstance, otherwise we would not be here. Interstellar clouds on the other hand, are invariably changing, evolving, because the forces pushing them this way or that are not balanced. Flux, not fixity, is the hallmark of matters astronomical. Sometimes peace is obtained only in star birth.

These important concepts have been mentioned so that the reader will have some feel for what astrophysicists are doing behind the scenes as they explore the nature of interstellar matter. The researcher snuggled up to a computer is probably "running" models which involve equations of balance. It all boils down to deciding or discovering which forces, energies, or processes to put into one's equations. The extraordinary thing is that astronomers have enough data on interstellar cloud sizes, masses, temperatures, and densities, as well as chemical composition, that the evolutionary processes in interstellar space can be accurately modelled.

Barnard suspected that there was dark matter between the stars. Others proved it was there, but it wasn't until astronomers began to make quantitative measurements of the amount of matter, solid and gaseous, that they could begin to do very much physics. How astronomers obtain the data that they use in their models will be the subject of the following chapters.

Note

[1] Quoted in Clark, R.W. (1984) *The survival of Charles Darwin*, Avon books. p. 9.

Part 4   Modern Aspects of Interstellar Matter

We finally touch on more etheric interstellar matters, studies of atomic and molecular gases, as well as magnetic fields between the stars. Knowledge about these constituents has exploded in recent decades and illustrates basic aspects of how astronomy has evolved into astrophysics. The study of interstellar matter has revealed what our Milky Way galaxy looks like. Studies of interstellar molecules within the dark clouds have given so much information about the physical and chemical conditions in clouds in space that processes of star formation have been clearly modelled for the first time. Much of what was once mystery has fallen victim to the relentless probing of astronomers using radio, optical, and infrared telescopes. In recent years even stellar birth pangs have been directly observed. Finally, questions relating to the origin of life must now be broadened because the answers may increasingly involve interstellar matters such as have been explored in this book.

CHAPTER 15

The Milky Way

There is a way on high, conspicuous in the clear heavens, called the Milky Way, brilliant with its own brightness.—Ovid[1]

The Milky Way Notion

Philosopher Immanuel Kant, way back in 1755, had it all figured out, and quite correctly. But if ever there was an idea ahead of its time it was his. The Milky Way's band of light that sweeps across our heavens had been commented upon since antiquity but no one knew why it took the shape it did, except Kant. Today we agree with him that it is a manifestation of the fact that we live inside a flattened, disk-shaped galaxy, and, being located amongst the welter of stars, we see millions upon millions fading off into the distant glow we call the Milky Way. When we look directly outward, away from this disk, we see relatively few stars and the heavens appear dark; and today we also know of countless other galaxies, some spiral shaped, others elliptical. The current model to account for the appearance of the Milky Way is well understood. (Note that the name *Milky Way* is given to both to describe the band of light we see with the naked eye and to the galaxy of which it is a manifestation.) But such a clear understanding was very long in coming.

However, in Kant's time his was a lone voice in the wilderness. In 1755 he wrote, in a *General Natural History and Theory of the Heavens: An Essay on the Constitution and Mechanical Origin of the Whole Universe Treated According to Newton's Own Principles*, that

it is astonishing that the observers of the heavens have not long since been moved by the character of this perceptibly distinctive zone [the Milky Way] in the heavens, to deduce from it special determinations regarding the positions and distribution of the fixed stars.[2]

Kant inferred from logical arguments that the "nebulous stars" were likely to be great star clouds at vast distances and that the visible Milky Way was a manifestation of the "island universe" in which we lived. The nebulous stars referred to cloudy objects seen through ancient telescopes. In 1610 Galileo had already suggested that the nebulae were groups of stars "distributed in a wonderful way."[3] But how could Kant really prove that the nebulae were other galaxies? Inference was not enough.

A century earlier, in 1656, Huygens had observed the Orion nebula and in his *Systema Saturnum* described it as a "portent to which I have been unable to see anything similar among the fixed stars"[4]; and so it remains to this day, one of the most beautiful of all the nebulae (Figure 1.5). It can be seen with the naked eye on a clear (northern) winter's night as the fuzzy object which is the central "star" in the sword of Orion. The sword hangs from Orion's belt, three equally bright stars across the middle of the constellation. (Southern hemisphere observers see the sword point upward relative to the horizon!)

The nebulae were to gain greater fame when in 1781 Charles Messier listed 103 of them in a table of objects he believed should be avoided by comet seekers. To this day the major nebulae go by their numbers in this catalog; thus the Andromeda galaxy is known as M31.

Herschel Again

In 1790 William Herschel changed his mind about the stellar nature of the nebulae when he discovered a planetary nebula (Figure 15.1), a diffuse nebulosity symmetrically located about a star, which destroyed his belief that nebulae were composed of distant stars.[5] He and his sister Caroline had published a catalog of 1,000 nebulae and he had drawn attention to "the wonderful and inexplicable phenomenon observed around Orion."[6] But now the discovery of a planetary nebula convinced him of the existence of a widespread nebulosity mixed with the stars, which consisted of "a bright fluid of nature completely unknown to us."[7]

Sir William was the first of the great astronomers to seriously confront the nature of the Milky Way in his epochal effort to understand all there was to know about the universe. "The subject of the construction of the heavens . . . is so extensive and important a nature, that we cannot exert too much attention in our endeavors to throw all possible light on it" he wrote. He cautioned his contemporaries that in the pursuit of nature's secrets:

If we indulge a fanciful imagination and build worlds of our own, we must not wonder at our going wide from the path of truth and nature; but these will vanish like the Cartesian vortices, that soon gave way when better theories were offered. On the other hand, if we add observation to observation, without attempting to draw not only certain conclusions, but also conjectural views from them, we offend against the very end for which only observations ought to be made.[8]

He pressed forward, trying to determine the construction of the heavens, in retrospect an impossible task, since he had only a telescope with which to work. Astronomical photography had not yet been invented so

FIGURE 15.1. The planetary nebula (the Helix or NGC 7293) in Aquarius. When William Herschel first saw that such ring-shaped nebulae existed he was forced to change his mind about the widespread belief that all nebulae were groups of stars. The planetary nebulae clearly consisted of diffuse nebulosity surrounding a central star. (National Optical Astronomy Observatories)

he could only see that to which his eye aided by the telescope was sensitive.

Concerning a number of nebulae Herschel felt "a certain air of youth and vigor to many of the very regularly scattered regions of our sidereal stratum." On the other hand, "Some parts of our system indeed seem already to have sustained greater ravages of time than others, if this way of expressing myself may be allowed; for instance, in the body of Scorpio is an opening, or hole, which is probably owing to this cause."

This was one of the dark regions later photographed by Barnard. Apparently Herschel thought it exhibited the ravages of time—wear and tear, one might say.

He pointed out that at the boundary of one such hole is one of the "richest and most compressed clusters of small stars I remember to have seen . . . and would almost authorize a suspicion that the stars, of which it is composed, were collected from that place, and had left a vacancy."

This was 1817, his first mention of holes in the heavens, the concept that was later to interfere with Barnard's understanding of the dark markings.

Herschel immodestly believed that his work on the Milky Way would never be surpassed; at least this is the feeling we are left with upon reading a summary of his work:

What has been said of the the extent and condition of the Milky Way in several of my papers on the construction of the heavens, with the addition of observations contained in this attempt to give a more correct idea of its profundity in space, will nearly contain all the general knowledge we *can ever have* of this magnificent collection of stars.[9] [my italics.]

Herschel went on to catalog the nebulae, and in 1864 *The General Catalog of Nebulae and Clusters* was published by him and his son, John. Ninety percent of its contents had been discovered by them. The New General Catalog of 7,840 nebulae was subsequently published in 1888 by John Louis Dreyer and its NGC designation for nebulae and galaxies is widely used today. (A supplemental *Index Catalog—IC—of* 5,086 more galaxies and assorted other nebulae was published in 1894 and both were reprinted by the Royal Astronomical Society in 1953. These remain the astronomer's basic catalogs for nebulae and galaxies.)

The discovery that the nebulae were often distant galaxies was to elude the Herschels, and William never made the connection between the Milky Way and its possible relationship to the spiral nebulae. However, William Herschel had suggested that our local starry realm was elongated in the direction of the Milky Way and presented an irregular outline,[10] but his son John is credited with being the first astronomer to suggest that the elongation of the stellar disk is due to rotation of the system.[11]

The Milky Way as a Thing Apart

David Gill, in 1891, who later became Sir David Gill, Her Majesty's Astronomer at the Cape, speculated on the issue of whether the Milky Way was a separate system of stars. He was aware that the Milky Way stars were bluer and whiter than the naked eye stars. "Therefore," he wrote, "we may come to the very remarkable conclusion that the Milky Way is a thing apart, and that it has been developed perhaps in a different manner, or more probably at a different and probably a later epoch from the rest of the sidereal universe." The sidereal universe he referred to was taken to be the star-filled heavens visible to the eye. The difference in color of the Milky Way stars set them apart, Gill thought, which they do, for it is now known that they are generally younger and hotter than the others, for reasons he could not guess at. (The stars in the Milky Way have most recently formed and are known as Population I objects, while the stars well away from the Milky Way, as well as in globular clusters, and far older and redder, are known as Population II objects.)

In 1900 Cornelis Easton attempted to explain the distribution of stars in the sky in terms of a picture of what the Milky Way might look like from the outside. He concluded that it was a ring, not necessarily with the sun at the center, and wondered whether the sun might be located in this ring. Stellar accumulations in different directions were probably at different distances from the sun.

There is little reason to hope that the great problem of the constitution of the visible universe will be definitely solved in the near future.[12]

Easton's glorious understatement may still apply, depending on which astronomer you ask.

In 1913 this same man made a bold attempt to infer the spiral structure of the Milky Way by interpreting the distribution of stars seen against the sky. The diagram he derived showed the spiral arms curving in the opposite direction to what is now known to be the case, and the galactic center was placed in the constellation Cygnus, nearly a quarter of the way around the sky from today's known center in Sagittarius. His paper included a masterful drawing of the distribution of stars in the sky, with the dark regions (the vacancies) in Cygnus beautifully evident. Easton's efforts were laudable but premature. However, he added a disclaimer, "I wish to insist upon the fact that the sketch does not pretend to give an even approximate representation of the Milky Way."[13] How often do modern researchers dare to be as honest?

Myriads of Stars

It was Edward Barnard, late in the nineteenth century who had first seen the Milky Way in time exposure photographs which, as he put it, revealed "myriads of stars" against which he sometimes saw the dark markings.

As we have described in earlier chapters, our terrestrial vision of the heavens is clouded by fog that drifts between the stars, a fog of interstellar dust that prevents light from the outskirts of the Milky Way from reaching the earth. This phenomenon creates the "zone of avoidance" (chapter 8), a band centered along the Milky Way in which no distant galaxies can be seen. Optical astronomers are blinded to what lies beyond the dust and are like travellers lost in a forest, whose gaze meets only trees, no matter where they turn their heads.

By the early twentieth century, however, despite the blinding effects of the dust, astronomers were at last seriously struggling to explain the existence of the countless stars in the Milky Way, and had begun to reconsider the notion that there might be other "island universes" beyond the stars. The ghost of Kant had returned to haunt the observatories. But astronomers are a cautious breed, and they struggled for decades as the importance of the concept grew in their minds.

The awareness that the Milky Way's band of stars was a manifestation of our living inside a galaxy that might be like other spiral "nebulae" clearly belongs to this century. Detailed attempts to determine its structure are barely fifty years old. The myriad stars revealed to telescopes pointed along the Milky Way could only be explained by recognizing that the local swarm of stars was distributed in some vast flattened volume of space, so that when seen along the axis of flattening so many stars are seen that their faint light blends in a diffuse glow.

The Conceptual Leap

The breakthrough in understanding the shape of the Milky Way galaxy came from the work of Harlow Shapley, who began his career as a crime reporter for a small-town newspaper in Kansas covering dramatic stories about fights between drunken oilmen.[14] Shapley determined the distances to various types of variable stars, to globular clusters—which exist well away from the Milky Way—and open clusters, which exist in the galactic disk, in the zone of avoidance. His work showed that the globular clusters filled a volume of space far larger than had been believed before, a volume which quite naturally enclosed the plane of the Milky Way as a disk of stars. He concluded that

all known sidereal objects become a part of a single enormous unit, in which the globular clusters and Magellanic Clouds, these extensive and massive systems, are clearly subordinate factors.[15]

He recognized that the system's equatorial region was devoid of globular clusters and estimated its width to be 3000 to 4000 parsecs, less than $1/25$ of its probable extent in the galactic plane. The width estimate was about ten times as great as the currently accepted value, but otherwise his model was very close to the modern view. The center of this system (the Galaxy) was in the direction of the rich star clouds of the Milky Way and planetary nebulae, diffuse nebulae, open clusters, the naked-eye stars, and all the stars in anyone's catalog all lay in the plane of this system. However, he was wrong in his original conclusion that "it appears *unlikely* that the spiral nebulae can be considered separate galaxies of stars" [my italics].

Yet Shapley was only partially correct, because the spiral nebulae did turn out to be separate galaxies of stars. At the time Curtis said so and he and Shapley were subsequently involved in a dramatic debate, held on April 26, 1920, before the National Academy of Sciences, in which the nature of the nebulae was argued. The matter was put to rest in December 1924, when Edwin Hubble announced that he had found the spiral nebulae to be very distant, based on his observations of variable stars in objects such as M31. At last astronomers were certain that the Milky Way was but one of countless galaxies in the universe.

George Abell has suggested that

the realization, only a few decades ago, that our galaxy is not unique and central in the universe ranks with the acceptance of the Copernican system as one of the great advances in cosmological thought.[16]

If we were to ask the well-informed-person-in-the-street about an opinion on this subject, I suspect that he might be aware of the Copernican breakthrough in regard to the sun's location in the universe, but they would stare blankly when asked about the nature of galaxies. It is widely taken for granted that we live in one of countless examples, but few realize how difficult it was for astronomers to come to terms with this fact. (That struggle is well documented elsewhere.[17]) Perhaps the reason for the widespread ignorance of this interesting stage in astronomy is due to the lack of good public relations for the subject, publicity amply afforded the Copernican-Galilean breakthrough by the Catholic Church's interest in the subject.

The proponents of the correct view, such as Curtis, did not have to break out of a millennia-old and widely held belief about the nature of the nebulae. There was nothing in the Bible to stand in the way. The conceptual leap, while of great interest to astronomers, was of little concern to those whose lives hinged around more ecclesiastical matters. Therefore none of the astronomers had to confront devout zealots with millennia-old beliefs about the uniqueness of the Milky Way, a lobby that would surely have done much more to bring the matter to the attention of the media, and thus the public.

Dust Gets in the Way

Kapteyn, who we have mentioned earlier, was one of the astronomers who realized that unknown absorption effects would produce incorrect distances to stars. This would influence Shapley's results concerning the size of the Galaxy. Kapteyn spent much of his life pursuing a huge observation program to determine the correct structure of what was in his day still called "the universe," the Milky Way galaxy. Despite Kapteyn's initial expectations about the existence of interstellar matter, he surprisingly concluded that it did not exist. This implied that the decrease in stellar densities he observed in all directions around the sun was not due to absorption of light, but represented a real decrease in the number of stars. This, in turn, forced him to believe in the concept that the sun was at, or near, the center of "the universe."

In a letter written to George Ellery Hale in 1915 he struggled with this problem and wrote:

One of the startling consequences is that we have to admit that our solar system must be in or near the center of the universe, or at least at some local center.

Twenty years ago this would have made me skeptical. . . . Now it is not so— Seeliger, Schwarzschild, Eddington, and myself have found that the number of stars is greater near the sun. I have sometimes felt uneasy in my mind about this result, because in the derivation the consideration of scattering of light in space has been neglected. Still it appears more and more that the scattering must be too small and somewhat different in character from what must explain the change in apparent density. The change is therefore pretty surely real.[18]

By 1922 Kapteyn had enough data to publish his view of the way the universe looked, a view which came to be called "Kapteyn's Universe." This model lasted only as long as the idea of the lack of absorption was accepted. However, it served as a useful basis for further argument and discussion.

Kapteyn's "universe" was supposed to be ellipsoidal with a 5:1 axial ratio and with the major axis being about 16 kpc long. The center was located in a direction about 100° removed from the modern view (at old galactic longitude 77° and latitude −3°). The sun was only 650 pc from the center, and because this placed it in a favored position many doubts were raised as to the trustworthiness of the model. Yet Kapteyn could find no direct evidence for absorption which would force him out of his belief or conclusions. Across the Atlantic, however, Barnard's work had revealed the existence of large numbers of dark clouds, and many of these were located in the Milky Way. We may ask, with hindsight, why such information was not taken into consideration by Kapteyn. His quantitative efforts at measuring the effect of the obscuring matter were producing a negative result, yet the evidence for the existence of dark matter was already considerable, both as evidenced by both Barnard's photographs and Wolf's star counts.

This dichotomy highlights an important point regarding the distribution of interstellar matter. Barnard was photographing discrete "clouds" of the stuff, but the existence of "general" absorption was far more difficult to prove. Today it is taken for granted that clouds of interstellar dust are concentrations of material which is otherwise thinly spread everywhere between the stars in the Milky Way, but to reach this apparently simple conclusion took years of hard labor and many reputations were laid on the line in the quest.

The apparent lack of general absorption was confirmed by Shapley's work on star clusters, although he only observed globular clusters well away from the Milky Way and correctly concluded that they suffered no significant amounts of absorption.[19] However, the culprit that was leading Kapteyn to his incorrect conclusions, interstellar dust, lay in a region confined to the Milky Way and did not interfere with Shapley's observations of globular clusters situated well above or below the Galaxy.

A sense of how the obscuring dust might affect such observations can be had in the dramatic view of the outskirts of the galaxy NGC 253 shown in Figure 15.2. Imagine being trapped inside this galaxy and from that vantage point trying to infer its shape. The situation for us on earth,

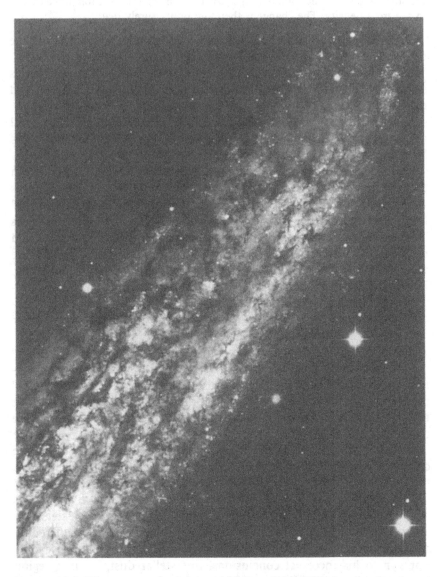

FIGURE 15.2. The outer parts of the spiral galaxy NGC 253, showing the elegant chaos created by the intermingling of extensive dust clouds and young stars that illuminate the matter. Older stars are widely scattered throughout the galaxy. Imagine the difficulty an observer in NGC 253 would have in attempting to see through the dust patches. The bright star-like images in the photograph are stars in our galaxy. (Copyright Anglo-Australian Telescope Board)

orbiting a star on the outskirts of the Milky Way, is little different from the fate of the extraterrestrials in NGC 253.

Jan Oort, a new star on the Dutch astronomical scene, was the first to derive the distance to the galactic center, and his conclusion was very different from Kapteyn's. He wrote:

The most probable explanation of the decrease in density [of stars] in the galactic plane indicated for larger distances is [that it is] mainly due to obscuration by dark matter. Such a hypothesis receives considerable support for the marked avoidance of the galactic plane by the globular clusters, a phenomenon for which up to this time no other well defensible explanation has been put forward.[20]

Oort believed this, but could not yet prove it. At this time, 1927, there was still opposition to the notion of interstellar dust, but by 1930 Trumpler's work and Eddington's battle cries had sealed the case. Trumpler found that the more distant clusters appeared to have larger diameters than expected, because their distances had been overestimated. The point is that if light is absorbed by interstellar matter, the distant clusters would appear fainter than they would otherwise, and thus astronomers incorrectly inferred that the clusters were farther away than they truly were.

Trumpler pointed out that the lack of reddening in globular clusters was the result of the fact that the absorbing medium was concentrated to the galactic plane, just like the open clusters themselves. He struck the astronomical nail right on the head when he wrote:

Perhaps this absorbing medium is related to interstellar calcium or to diffuse nebulae which are also strongly concentrated to the galactic plane.[21]

The scene was now set for the modern era, in which the study of interstellar matter was to lead to a definitive mapping of the Milky Way, a mapping which revealed that it is a spiral galaxy.

Notes

[1] Ovid, *Metamorphoses*, Book. 1, lines 168–169.

[2] Quoted by John Mood, *Astronomy Magazine*, February, 1988: 28.

[3] Abetti, G., Hack, M. (1959) *Nebulae and galaxies*, New York: Thomas Y. Crowell; quoting from "Sidereus Nuncius."

[4] Ibid.

[5] Berendzen, R., Hart, R., Seeley, D. (1976) *Man discovers the galaxies*. New York. Science History Publications.

[6] Ibid.

[7] Ibid.

[8] Herschel, W. (1817) *Phil. Trans*: 302–331.

[9] Ibid.

[10] Ibid

[11] Ibid.

[12] Easton, C. (1900) "A new theory of the Milky Way." *Astrophys. J.* 12: 136.

[13] Easton, C. (1913) "A photographic chart of the Milky Way and the spiral theory of the galactic system." *Astrophys. J.* 37: 105.

[14] Berendzen, et al., op. cit., p. 35.

[15] Shapley, H. (1917) "Globular clusters and the structure of the galactic system." *Publ. Astr. Soc. Pac.*: 42.

[16] Abell, G. (1982) *Exploration of the universe*, 4th ed., Saunders College Publishing, New York: 598.

[17] Berendzen, et al., op. cit., p. 35.

[18] Quote from a private communication between Kapteyn and Hale as given in Berendzen, et al. p. 75.

[19] Shapley, H. (1917) "Studies based on the colors and magnitudes in stellar clusters, first part: The general problem." *Astrophys. J.* 45: 130.

[20] Oort, J.H. (1927) "Observational evidence confirming lindblad's hypothesis of a rotation of the galactic system". *Bull. Astr*, Inst. Neth. 3: 281.

[21] Trumpler, R. (1930) "Preliminary results of the distances, dimensions, and space dstribution of open star clusters," *Lick Obs. Bull.* 14: 154.

CHAPTER 16

Radio Astrophysics

The Modern Study of Interstellar Matter

To appreciate the majesty and drama of interstellar processes and the nature of the Milky Way we have to look beyond what the photographs show, beyond anything our greatest telescopes reveal to the eye. We have to explore in our imagination. Yet our imagination must always be guided by the data that have been gathered by those telescopes, data that are invariably processed in computers, and then carefully and sometimes colorfully displayed in ways that make sense to the human mind. Enough is known about physics and chemistry to reveal that the laws determining interstellar processes are universal, and our understanding is ultimately related to what is learned in the laboratory.

The study of the gaseous component of interstellar matter, the focus of this section, is essentially modern, especially if our criterion is one of the explosion of knowledge. Observations of "spectral lines" produced by this matter reveal the secrets of interstellar gas clouds—their distance, temperature, size, density, and derived parameters concerning stability or evolution. Once the physical properties of the interstellar clouds are known, these parameters can be "plugged" into computer programs that model the processes of cloud evolution and star formation.

In order to communicate a sense of what the astrophysicist does to obtain information about this interstellar gas, the radio observations of the most basic atom in the universe, hydrogen, which has a unique radio signature, will be highlighted. The concepts described may be applied across a wide range of astronomical studies and represent the backbone of what modern astrophysics is about. They reveal how observations of spectral lines give information about the interstellar gas—atomic and molecular—that accompanies dust grains in their endless vigil between the stars.

Invitation to the Party

It was a cold January day in 1961 when I saw my first radio telescope, the awesome 250-foot diameter steel monolith at Jodrell Bank, which dwarfed everything else in the Cheshire countryside (Figure 16.1). The

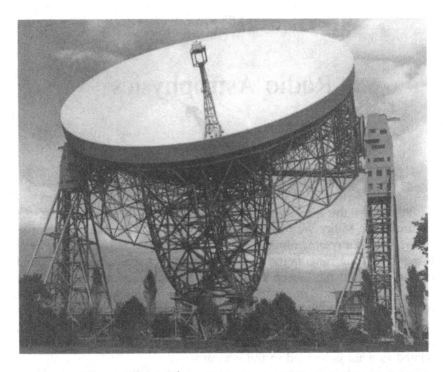

FIGURE 16.1. The Mark IA or Lovell Radio Telescope, at Jodrell Bank, England. This is the structurally strengthened version of the telescope used by the author in his early studies of interstellar matter from the laboratory high in the support tower on the right. (Nuffield Radio Astronomy Laboratory)

damp grayness clung to me as drizzle hung in the air, reluctant to touch the earth. The enormous steel radio reflector was held 150 feet above the ground by support towers which trundled in unison around a circular track, driven by an ancient analog computer to follow celestial objects. At the time this was the world's large radio telescope and the sense of immensity that it imparted to me was only equaled years later by a visit to Yosemite Valley in California.

I had been invited to come to Jodrell Bank to join in the exploration of interstellar matter. For a little over a decade astronomers had been using radio telescopes, and this was a wonderful time to be entering a new science. I didn't know it then, but the astronomical heavens were about to be laid bare by new generations of telescopes, not just giant radio reflectors, but satellite- and balloon-borne telescopes capable of studying ultraviolet, infrared, and x-ray emission from cosmic objects.

From the start that giant telescope symbolized for me the embodiment of the human spirit and its need to expand our boundaries of awareness, of consciousness, to go beyond the limitations of our senses. The next fifteen years were to be a period of discovery unprecedented in astronomy and would change our perception of the nature of the universe and its contents. This was especially true in the realm of interstellar matter.

Thus began my astronomical career, an adventure which was to bring to me the thrill of discovery mixed with endless hours of fruitless study of faint radio whispers from space, which did not always proclaim what we hoped to find.

I had been specifically hired to work in a group of graduate students (led by Rodney D. Davies) to study interstellar hydrogen. My baptism came quickly when I was handed a piece of paper on which was drawn a peculiarly shaped line.

"I'd like you to analyze this," my thesis advisor said.

I stared in disbelief. I recall that the diagram looked somewhat like the one shown in Figure 16.2, although no one had labelled it for me. (The meaning of these labels will be discussed later.) What was there to analyze, I wondered? This was just a squiggly line on a piece of paper!

"This is a spectrum produced by interstellar hydrogen gas," Rod Davies went on. "See what you can learn from that profile."

I was to hear the term *profile* thousands of times during the next decades and the appellation is an apt description. You can recognize a friend by his profile. Demographers and census-takers construct profiles of a community. Profiles contain a lot of information. The hydrogen-line profile, a spectrum, shown in Figure 16.2 also contains a great deal of information.

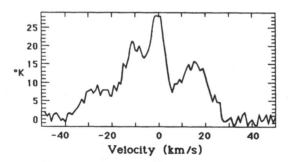

FIGURE 16.2. An idealized 21-cm spectrum due to interstellar hydrogen. This form of presenting the data shows the way the strength of the received radio signal (in units of antenna or brightness temperature, see text) varies with wavelength. Before plotting the data the wavelength shift due to the Doppler effect has been converted into a velocity. The zero velocity means that the gas is not moving with respect to the local standard of rest adopted by astronomers as their reference frame.

The Search for the Signature

Hydrogen is the most common element in the universe, and the road to study of radio waves from the interstellar variety of it—cold, neutral hydrogen (HI)—began in occupied Holland in 1944 when a young student, Henk van de Hulst, first derived the wavelength at which this gas might be observed.

He recalls[1] the day when he first calculated the wavelength on a piece of squared paper prepared for a presentation to the Dutch Astronomer's Club. He confesses that it is a sign of the times that graduate students may now be failed for not being able to perform the same calculation in an exam.

In 1944 the population of occupied Holland already knew the war was being won by the Allies; the only questions were when, and who would survive. The students were still at work, but some of their professors could only afford to make irregular appearances under conditions of considerable secrecy if they happened to be active in the resistance movement. Jan H. Oort was one such advisor who made dangerous forays to visit his students. It was he who encouraged van de Hulst to explore how radio waves from interstellar matter might be usefully studied after the war ended.

The nationwide concern over who would survive the expected invasion to liberate Holland was highlighted in van de Hulst's circle of acquaintances when the parents of a colleague from Arnhem wished that their son would return home. They thought that to be close to the sea was dangerous and wished he would return to the safety of his home. (The Allies launched their first effort to liberate Holland in a major, and unfortunately disastrous, airborne assault on Arnhem!)

van de Hulst had been invited to the University of Leiden to study there for a few month because his own professor at Utrecht had been interned by the occupation forces. At the observatory, where he resided, van de Hulst occasionally heard celebrations in the room above his, usually after Allied supply drops had been successfully collected by the members of resistance, one of whom was a technical assistant at the observatory.

The universities were officially closed during the occupation and so van de Hulst's formal work on his graduate degree had to wait. But on his own initiative he had begun to work on an overambitious thesis project in which he planned to cover everything there was to be said about interstellar matter. On one of his surreptitious visits Oort asked van de Hulst to give a talk about radio emissions from the universe and what they might tell astronomers about interstellar matter. He was more than eager to oblige, because this fitted perfectly with what he wanted to do for his thesis anyway.

Bear in mind that the concept *radio astronomy* did not yet exist, and that the Dutch scientists were isolated from astronomical research occurring elsewhere in the world. However, van de Hulst had available two pre-war papers, one by Grote Reber on the observations of the radio emission from the Milky Way, and a theoretical article attempting to explain the radiation.

In preparing his presentation he spent months working on the theory of the so-called free-free emission, produced by clouds of hot, ionized gas in which free electrons interact with each other and radiate energy that covers a broad spectrum of wavelengths (see below).

He also spent a few additional weeks on a slightly less difficult problem, that of the emission of spectral lines from the ionized matter as it undergoes a process known as *recombination*. When a hydrogen atom is ionized the electron formerly in the lowest possible orbit or energy level (the ground state) is either kicked completely out of contact with the proton, or it moves to an orbit far removed from the original level. The electron may select among a large number of orbits (orbitals), each of which represents a different total energy for the atom. In the process of recombination the electron falls to lower and lower levels until it is once more in the lowest, the so-called ground state, which involves the least amount of energy.

In the process of falling into lower energy levels the electron loses energy by radiating at a specific wavelength determined by the energy difference between the levels involved.[2] The recombination of hydrogen atoms produces a large number (hundreds) of spectral lines, each with a characteristic wavelength. These are broad, weak lines and are routinely observed from emission nebulae such as Orion and M16 (Figures 1.4 and 8.6, respectively).

The theory relating to free-free emission had been worked out in a classical manner (that is, without quantum mechanics) by Professor H. A. Kramers in Leiden, who showed how it could lead to the explanation of the *recombination lines* using quantum theory. van de Hulst found that he could calculate their wavelengths, but his work showed that they would be so wide that they would overlap and be unobservable. It turned out that he had made an error in a calculation, multiplying something by 100 instead of dividing, an error of the most common sort, which he never discovered until 1959 when others calculated the wavelengths correctly and recombination lines from ionized hydrogen in emission nebulae were discovered.

Finally, he spent only a couple of days on a third topic, the cold hydrogen (HI). He quickly concluded that the HI should produce a spectral line at a 21.2-cm wavelength (1,420.40575 MHz). He had no problem with this calculation and believes that the time was ripe for this derivation which was soon performed elsewhere by others oblivious of his work. To perform the calculation he only needed three formulae on the properties of atoms and spectral lines, all of which were readily available in different libraries in Leiden. While the books he needed were not within arm's reach, they were within easy walking distance, and so the task was soon completed.

He did have to make one estimate, of the strength of the so-called magnetic dipole in the hydrogen atom, to complete his calculation. His educated guess was right, and it was with a sigh of relief that his work proved correct years later when, one morning in the office of H.I. Ewen at Harvard, he was shown, spread out over the floor, chart records which revealed the detection of the radio waves from interstellar HI.

Ewen and E.M. Purcell, veterans of radar research during the war, working at Harvard, had on March 25, 1951, obtained the first signals from interstellar hydrogen. Seven weeks later, on May 11, C.A. Muller and J.H. Oort in Holland, also received the signals. They had suffered a serious fire in their equipment which caused a setback of many months. Posterity records that three groups, at Harvard, in Holland, and in Australia, simultaneously reported the detection of the 21-cm line, but the Harvard group was the first to record the signals.

Radio Observations of Interstellar Matter

To understand the nature of interstellar matter requires processing subtle bits of information and formulating theories as to why things are the way they are. The tools of astrophysics, a branch of science barely a century old, allow these data to be obtained and interpreted. Models representing the observations are then expressed as mathematical formulations which are processed in computers so that interstellar events can be speeded up in simulations that duplicate events in space. These models require information, for example, about the chemical constituents, temperature, density, and the sizes of clouds in space. The primary technique for studying the gaseous interstellar matter involves the observation of spectra, or profiles, such as seen in Figure 16.2, produced by radiation from atoms and molecules. This is known as radio spectroscopy.

Below we will describe radio spectroscopy in simple terms. It represents a close parallel to optical spectroscopy, which is usually discussed in basic astronomy textbooks and involves spreading light into its basic wavelengths—by using prisms, for example. In the radio domain the spectrum is studied by tuning the radio receiver across the range of wavelengths of interest. Also bear in mind that radio waves pass freely through space at the speed of light. Their wavelengths range from millimeters to tens of meters. They travel unaffected by interstellar dust because their wavelengths are so much longer than the size of the dust particles (10^{-5} cm). The tiny particles are essentially invisible to the radio waves.

Radio waves from space are generated in several ways. They may also be absorbed on their passage to the telescope by interstellar clouds of ionized gas, atoms, or molecules. The astronomer can recognize these various phenomena in the data.

Three processes have a relevance to events in interstellar space. The first two produce a wide spectrum (or continuum) of radiation and are called thermal and nonthermal emission. The third process generates narrow frequency, or spectral-line, radiation and is produced by atoms and molecules.

Figure 16.3A represents a swarm of particles inside a cloud of ionized matter consisting of rapidly moving electrons and slower moving ions (mostly protons). Motions of the particles are of the order of 10 to 100 km/s. When an electron is deflected from its motion in a straight line by interaction with a proton or another electron, it will radiate away electromagnetic energy as its trajectory is altered. Depending on the temperature of the gas, the largest amount of energy will be emitted as a radio wave, light, heat, or other form of electromagnetic radiation. The intensity of the observed radiation varies with wavelength over a broad spectrum, usually decreasing with increasing wavelength. By making measurements of the brightness of a radio source at various wavelengths the astronomer can recognize whether thermal emission is present, as for example in clouds of hot gas such as the Orion nebula (Figure 1.4).

The second process is known as nonthermal, or synchrotron, radiation, because it was first observed near particle accelerators known as synchrotrons. This radiation increases with increasing wavelength, a signature that is essentially opposite to that of thermal emission. Figure 16.3B represents the motion of a cosmic ray electron traveling near the speed of light as it encounters a magnetic field in space. The cosmic ray electron will spiral about a magnetic field line and in so doing radiate energy in the direction it is travelling at any instant along its curved path (the tangential direction). The energy of the electrons and the strength of the field determine the shape of the wideband spectrum of the emission and whether the peak of this radiation appears as light, radio waves, or x-rays, for example.

Spectral Line Radiation

The process we need to consider in order to interpret the profile in Figure 16.2 is *spectral line radiation,* and here we will explore the simplest case, that of the radiation from the hydrogen atom. Other spectral lines in radio astronomy may be regarded as being created in roughly analogous fashion.

The hydrogen atom at its coldest exists in the lowest energy (ground) state. Here the proton at the nucleus has an electron in orbit about it (Figure 16.4). Imagine that both the electron and proton spin, either clockwise or counterclockwise. When the two spins are in the same di-

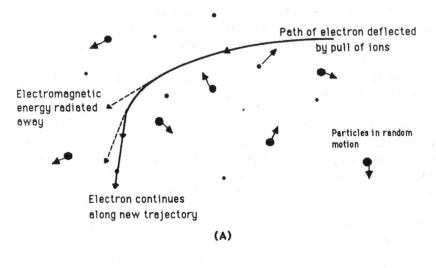

Electrons ✎ and ions ☌ moving about inside a hot gas cloud

Path of electron deflected
by pull of ions

Electromagnetic
energy radiated
away

Particles in random
motion

Electron continues
along new trajectory

(A)

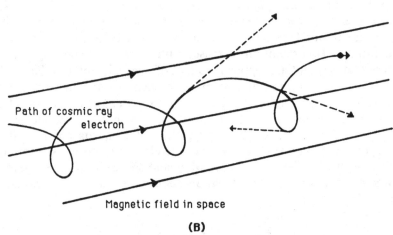

Path of cosmic ray
electron

Magnetic field in space

(B)

FIGURE 16.3. (A) The thermal or free-free emission process. Electrons and protons in an ionized gas move about freely. When an electron is deflected from its path by a near-collision it may radiate energy in the form of heat, light, or radio waves. (B) The nonthermal or synchrotron emission process, in which cosmic ray electrons, which permeate all of interstellar space spiral around magnetic fields and thereby radiate some of their energy in the form of radio waves. Energy is emitted in the direction of travel at any instant in the electron's path.

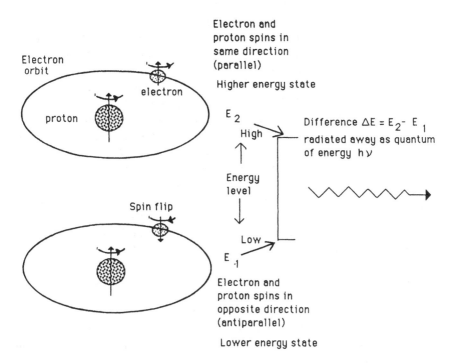

FIGURE 16.4. A schematic representation of the way in which a hydrogen atom radiates energy with a wavelength of 21-cm to produce the so-called hydrogen line from interstellar matter. See text.

rection (parallel) the atom contains an amount of energy slightly greater than when the two spins are in opposite directions (antiparallel). The atom thus has two possible energy levels in its ground state, with energies E_2 and E_1 as shown in the diagram. As the result of collisions an atom may change its spin state from parallel to antiparallel. This change in spin is called the spin-flip or *hyperfine transition*. The energy difference ($\Delta E = E_2 - E_1$) is radiated away as a radio wave at a 21.2-cm wavelength. This follows from Planck's law, footnote 2. The inverse process also acts; radio waves at a 21.2-cm wavelength may be absorbed when they strike the cloud (see below).

It is important to recognize that an atom or molecule contains a certain amount of energy which is made up of a variety of motions. In the case of the HI these motions are simple—orbital and spin. But for molecules the rotation of the molecule, end-over-end for example, or vibration of the atoms with respect to each other, contributes more energy. At issue is that whenever an atom or molecule changes its energy state for some physical reason, this difference in energy is either radiated away or absorbed as a quantum (a packet) of radiation, such as a light or radio wave with a very specific wavelength.

The loss of energy by radiation from HI atoms acts to cool the interstellar cloud slightly, but the cooling is more than balanced by energy gained from further absorption of radio waves and from cloud—cloud (hence particle—particle) collisions.

The natural emission produced by the hydrogen atom generates a "line" at a very specific frequency (or wavelength) whose natural width is infinitesimal (10^{-16} Hz). However, motions within an HI cloud in space introduce Doppler shifts,[3] so that the observed spectrum has a finite width determined by motion of the particles which, in turn, depends on the temperature in the gas. This is known as its *kinetic temperature*.

Figure 16.5 represents a perfect HI spectral line. The statistics of random motions within the cloud are such that the emission is centered about the "rest" frequency and decreases to either side. The shape of this curve is called a Gaussian or "normal" distribution familiar to statisticians and school teachers. The fact that the spectrum has a Gaussian shape means that most of the atoms in the hydrogen cloud being observed by the telescope appear to be moving across the line-of-sight and their average motion defines the central velocity of the cloud. A decreasing number of atoms have components of their motions toward or away from the observer, with the smallest number travelling either directly toward or directly away from us. The spread in velocity is directly related to the temperature of the gas in the cloud.

A 21-cm hydrogen line spectrum is displayed as intensity, often in terms of a temperature unit, related to agreed-upon calibration practices, versus velocity, because not only is the velocity a measure of the Doppler effect,

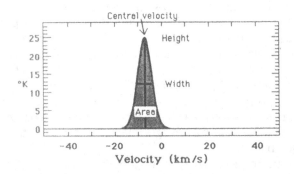

FIGURE 16.5. The basic Gaussian-shaped spectral line. The important parameters are its height, measured in units of temperature which are calibrated with respect to agreed upon international standards; the width; and the area under the profile, which is a measure of the number of atoms in the line-of-sight being observed by the radio telescope. The horizontal axis is most conveniently displayed as a Doppler shift with respect to the local standard of rest, an internationally agreed upon reference frame within which the Sun moves. The shift is expressed as a velocity and is related to frequency as indicated.

but this immediately provides the information of greatest interest. The vertical temperature scale shown here is, however, not related to the actual temperature in the cloud. This scale is presented as an "antenna temperature," which refers to how the brightness of the spectral line is seen by the "eye" of the radio telescope. It is then converted to a "brightness temperature", which is a measure of how much radio energy is actually given out by the cloud.

When a radio telescope is used to make narrow-bandwidth (i.e., high frequency resolution) observations of the radio emission from several interstellar hydrogen clouds a profile such as seen in Figure 16.2 will be produced. The nonsmooth nature of the profile in Figure 16.2 is due to "noise" to be discussed below. In any event, such profiles contain a wealth of information about the physical conditions of the gas responsible for the signals.

Variations on a Theme

The path of the radio energy from a cloud to the telescope and hence to the data display device is illustrated in Figure 16.6. HI atoms move randomly in the cloud and radiate 21-cm radio waves in all directions. They pass freely through space until some of them reach the radio telescope where they are focussed, amplified, and processed with a multichannel radio receiver known as a spectrometer. This produces the idealized profile (Figure 16.5) for study. Such a receiver may contain from 1,000 to 2,000 separate frequency channels which simultaneously produce data.

Figure 16.7 shows the consequence of mass motion of the cloud with respect to the observer. A distant cloud at rest (that is, having no motion toward or away from the observer) will emit a 21-cm spectral line that is observed at the predicted wavelength for the spectral line as derived from laboratory measurements. We say that this cloud has zero velocity.

If the cloud is moving away from the earth, the waves will appear to be stretched by this motion due to the Doppler effect and will thus be received at a slightly different wavelength. This Doppler shift is initially measured as a frequency shift because the technology of radio receivers involves frequency. Astronomers then use computers to convert the Doppler shift into a velocity of the gas cloud with respect to the observer. In so doing they have to take into account the earth's rotation and motion around the sun, which add Doppler shifts. The net result of all this data processing is to label the horizontal scale of spectra such as was shown in Figure 16.2 in terms of a velocity. (The Doppler shift of a cloud moving away from the observer will cause the spectral line to be shifted to positive velocities. In the case of a cloud moving toward the earth the Doppler shift produces a negative velocity.)

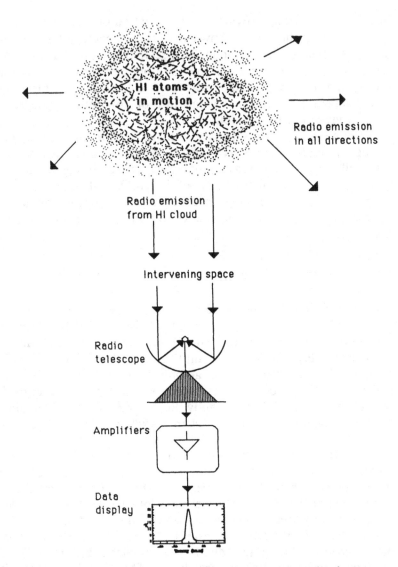

FIGURE 16.6. Hydrogen atoms in an interstellar cloud move randomly at a velocity determined by their temperature and radiate 21-cm waves which are gathered by a radio telescope, focussed, amplified, and displayed as a spectrum.

Observations of the central velocity of the HI spectrum gives information on the average motion of the distant cloud.[4] A cloud moving exactly across the line-of-sight will generate no Doppler shift (unless it travels near the speed of light, a condition which need not concern us here).

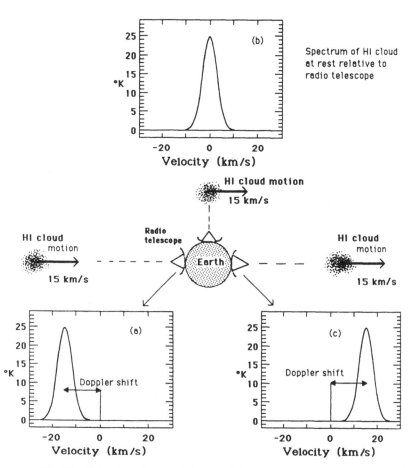

FIGURE 16.7. The Doppler effect and how it influences the 21-cm spectral line received by a radio telescope. We imagine that clouds of neutral hydrogen are moving at 15 km/s through space. In the upper example the radio signals are travelling toward the earth at right angles to the movement of the cloud. No Doppler effect is produced and the hydrogen spectrum is observed as centered at zero velocity. When the cloud appears to be moving toward the observer (left side) the spectral line is shifted to negative velocity, while if it is moving away from the observer (right side) it is shifted to positive velocity. In practice the telescope is moving with respect to the hydrogen clouds as a result of both the earth's rotation and its movement through space around the sun and with the sun. These motions are automatically corrected by the computer, so that only the Doppler shift of the HI cloud remains in the final presentation of the data.

The measurements of Doppler shifts in terms of a velocity is of extraordinary importance throughout astronomy. This allowed the recession of galaxies to be recognized, which then gave rise to the Big Bang model of the universe. Also, the study of the way interstellar clouds appear

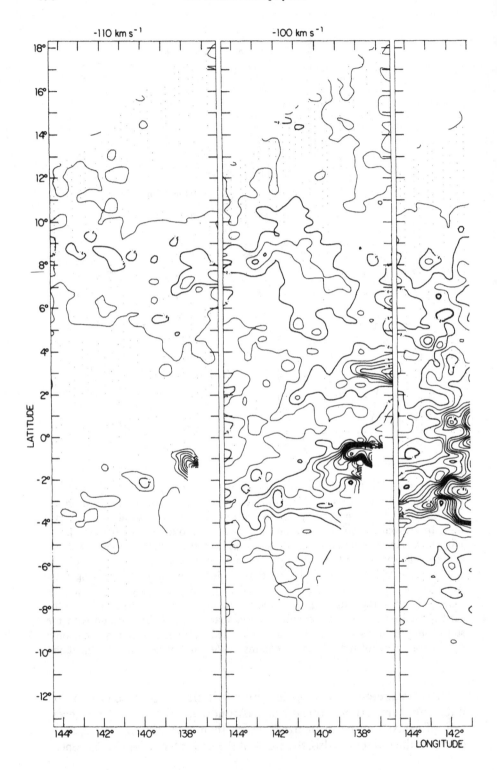

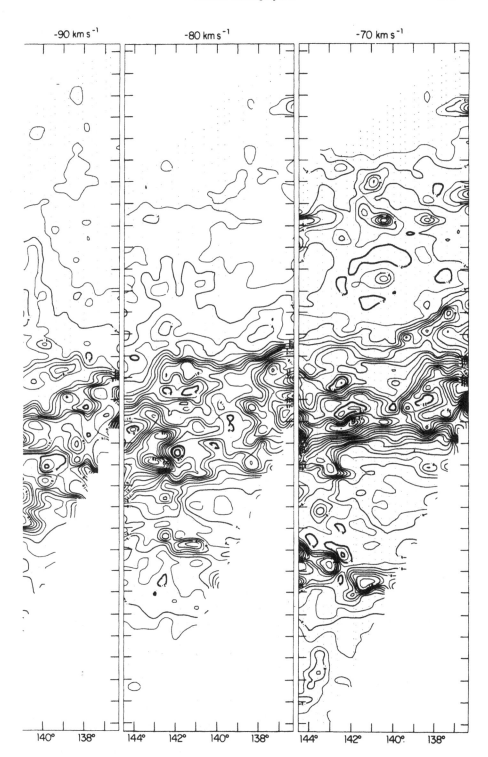

to be moving with respect to earth give the raw data which allow the structure of the Milky Way to be determined.

Radio Astrophysics

The width of a spectral line such as shown in Figure 16.5, measured at half peak intensity, can be related to the kinetic temperature of the gas, although line broadening due to other motions such as turbulence are sometimes observed. The area under the spectral line is a measure of how many atoms are involved in producing the spectrum. The area is initially given in units of brightness temperature times velocity, which can be converted into the number of atoms per square centimeter in the direction of the gas cloud (along the line-of-sight) using physical theory. In the case of a complex emission spectrum such as Figure 16.2, the astronomer has to determine how many individual Gaussian line shapes are present, sometimes a forbidding task.

If the depth of the cloud is known it becomes possible to determine the density of HI present, obtained by dividing the total number of atoms in the line of sight, which is given in units of area it turns out, by the depth of the cloud. The depth is usually found by assuming that the cloud is as deep as it is wide. Provided the distance to the gas is known, the angular width of the HI as mapped on the sky can then be converted to a linear depth by simple trigonometry. Then, if the total volume of the cloud is known, the total amount of gas can be found. Typical interstellar HI clouds have temperatures from 20 to 100 K, densities from 1 to 100 cm^{-3}, diameters from one to hundreds of parsecs, and masses from a few to tens of thousands of solar masses.[5]

The cloud size is found by mapping the spectral line over an area of sky, since the radio telescope accepts radio signals from only one direction at a time. When an area is mapped the cloud boundaries may be identified, as for example in Figure 16.8. This shows a large-scale view of the HI structure at a number of discrete Doppler velocities in an area crossing

---

FIGURE 16.8. A sample of hydrogen contour maps obtained in the direction of the constellation Perseus. The galactic coordinates of longitude measured from the galactic center, and latitude measured with respect to the mid-plane of the Milky Way, are shown. Each frame represents the intensity of the 21-cm line emission at a specific velocity. The band across the center in the right frame is due to hydrogen in the Milky Way. Negative velocity gas in this part of the sky is very distant, up to 15 kpc from the sun, and is mostly above the Milky Way (latitude 0°). (See chapter 18.) Sets of closed contours indicate the presence of HI clouds (e.g., upper part of right-hand frame). The lower right of each frame is an area not observed with the 300-foot telescope used to make the maps. (National Radio Astronomy Observatory)

the Milky Way in Perseus. Individual clouds are identified as sets of closed contours which may have a variety of shapes. The complexity of such maps gives an indication of why astronomers busy themselves for long periods of time trying to understand the details.

The interstellar hydrogen giving rise to the data in Figure 16.8 is about 6 kpc distant. Such maps may be compared with photographs of, say, interstellar dust in that region, or maps of the spectral line radiation from other molecules. The density and the mass of the cloud are found and then compared for clouds of different types, for various molecules, and in different directions in the sky. Once densities, temperatures, and cloud masses are known, the astrophysicist can begin to make models and explore possible evolutionary futures for the clouds. For the case of interstellar neutral hydrogen, the interpretation of the data is relatively straightforward, but when the radiation from molecules is studied, other factors, including chemical processes in the clouds, have to be taken into account (see below). So the modern study of interstellar matter proceeds.

Absorption Lines

Figure 16.9 shows what happens when the 21-cm line is observed in the direction of a radio source, say a distant supernova remnant. In this example, four clouds are lined up in depth through space, perhaps in four distant spiral arms. The supernova remnant is located between the two most distant arms. In the absence of the radio source the telescope would have seen the emission spectrum shown in Figure 16.2. But the radio source introduces its own signal at the telescope and its radiation covers a very broad range of wavelengths. Now, however, the radiation is absorbed by the three clouds in front of the source and this produces the absorption spectrum shown in the lower part of Figure 16.9. Close comparison with Figure 16.2 shows that an absorption dip is seen at the same velocity as each of the emission peaks that would have been present if the radio source was not there to confuse things. But only the three clouds in front of the source produce absorption. Note that the hydrogen gas absorbs at the same frequency at which it otherwise emits radiation. In this case the spin-flip transition would proceed the other way and absorb energy.

The 21-cm radio signal from the HI cloud beyond the radio source will still be seen in emission, just as it would be in the absence of the source. Determination of the velocity at which the absorption spectrum ends often allows the distance of the radio source to be found, provided a relationship between velocity and distance is known, which is true for a rotating galaxy (see chapter 17).

To sum up, measurements of width, depth, and area of emission and absorption line data, combined with mapping studies, give information

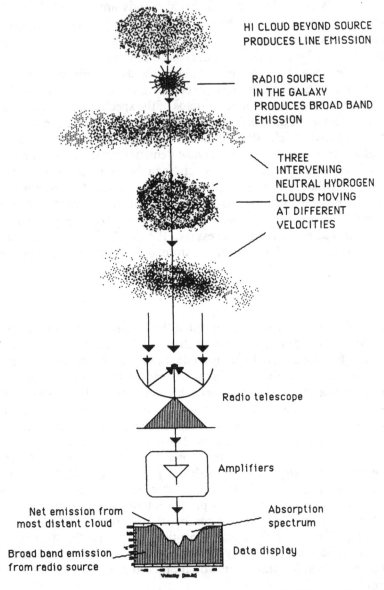

HI CLOUD BEYOND SOURCE
PRODUCES LINE EMISSION

RADIO SOURCE
IN THE GALAXY
PRODUCES BROAD BAND
EMISSION

THREE
INTERVENING
NEUTRAL HYDROGEN
CLOUDS MOVING
AT DIFFERENT
VELOCITIES

Radio telescope

Amplifiers

Net emission from
most distant cloud

Absorption
spectrum

Broad band emission
from radio source

Data display

Velocity [km/s]

FIGURE 16.9. The production of an absorption spectrum. A distant radio source, such as a supernova remnant, is seen behind three clouds of HI that are moving at different velocities with respect to the sun. The radio source produces a broad continuum of radiation at all wavelengths, while each of the HI clouds will absorb radiation at 21-cm but shifted by small amounts due to the Doppler effect. The net result is that the observed spectrum is seen in absorption, with the contribution of each absorbing cloud overlapping that from the other. The fourth HI cloud, behind the radio source, will be seen in emission, as is shown.

about physical conditions in the interstellar HI clouds. Analogous studies of molecular line data give information about the physics inside clouds of interstellar molecules which, in turn, allows their chemistry to be understood (chapters 18 to 20).

The Harsh Realities of Noise

The examples above showed clean-looking emission profiles. But radio observations are subject to *noise* (random energy) produced by the radio receiver, by ground radiation leaking into the antenna, and by the sky itself. Noise manifests itself as small, random changes in the temperatures displayed by the output of the receiver. When the noise level is high it is difficult to "see" the radio signal in the noise. This is like talking to someone at a noisy party. If the background noise is high it is very difficult to hear your companion, even if he is shouting. If the background noise is reduced, perhaps by asking people not to talk so loudly, it becomes easier to hear. The same is true in radio astronomy. If the background noise, most of which is produced by the receiver, can be reduced, cosmic radio signals are easier to detect.

Another convenient way exists to reduce noise in the data. Since noisy signals are random, the process of averaging a lot of observations reduces the amplitude of the noisy component, while the constant underlying signal remains unchanged. The noise decreases as the square root of the observing time. This is shown in Figure 16.10. The same spectral line has been observed for different lengths of time and the data averaged. The top example has 30 units of noise present. Let us assume that this was obtained from a one second observation, known as an integration. The next plot has 20 units of noise obtained by averaging two seconds of data. The next diagram has only 4 units of noise, which required a 56-second integration. Now the presence of the four spectral lines becomes fairly clear, whereas they were indistinguishable in the 1-second observation. The bottom diagram shows just 1 unit of noise obtained after 900 seconds (15 minutes) of integration. Now we clearly recognize the existence of four emission peaks.

Radio observatories spend tremendous amounts of engineering effort to reduce receiver noise to allow more sensitive measurements to be made in a given time. Also, it is not uncommon for astronomers to integrate on a source (that is, constantly track it and average all the data) for up to 50 or 100 hours in search of weak spectral lines.

And So On and So Forth

Variations on the themes discussed in this chapter exist in other bands of the electromagnetic spectrum, e.g., at light, infrared and ultraviolet wavelengths. In each domain the technical details of telescope and spec-

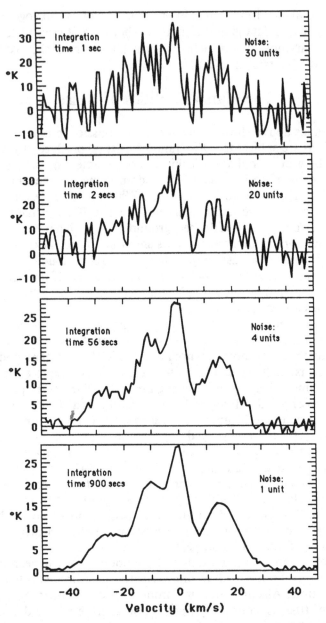

FIGURE 16.10. The influence of "noise" in the observations. The idealized spectral lines shown in previous diagrams tell only part of the story. In reality, electronic "noise" due to random motions of electrons in the electronic circuitry, the antenna, and the sky introduce a random component in the signal being observed. The level of the noise obtained on a spectrum depends on the duration of the observation—that is, how long an integration time is used. In this example, several integration times were used to observe the idealized spectral line shown

trometer design will vary and methods of data display and analysis will be slightly different. But at root lie the same goals; to extract from the observed spectra information on the physical conditions in the region giving rise to the emission or absorption lines.

Notes

[1] H.C. van de Hulst recently published a personal memoir in the Dutch popular astronomy magazine, *Zenit*, kindly given to me by Prof. van de Hulst.

[2] The frequency of the radiation is determined by the famous equation, $E = hf$, where E is the energy difference between the orbits, h is Planck's constant, and f the frequency of the radiation which, in turn, is related to the wavelength λ by $c = f\lambda$, where c is the speed of light.

[3] The shift in observed wavelength due to the motion of the object with respect to the observer, first studied by Austrian physicist Christian Doppler in 1842.

[4] A 1 km/s motion will generate a 4.74 kHz Doppler shift.

[5] Astronomical masses as usually reckoned in units of the sun's mass, which is 2×10^{33} grams.

in Figure 16.2. Each spectrum has had noise added to an identical underlying and noise-free spectrum. The top frame shows the noisy signal after only one second of integration. The noise level decreases as integration time is increased (by the square root of the time), until in the lower frame a 900-second integration has reduced the noise level by 30 times. Clearly the high-noise examples at the top prevent us from interpreting the correct shape of the profile. For this reason astronomers spend a great deal of effort and integration time to reduce the noise levels in their data. Only then can they be sure of what they have detected.

CHAPTER 17

The Shape of the Milky Way[1]

Seeing the Galaxy

Imagine standing on a planet in the Large Magellanic Cloud (Figure 1.1). In its night sky you would see millions of galactic stars splashed in a giant spiral swirl covering an area as large as a constellation. But this is only in your imagination because there is no way in which we will be able to book a flight to the Magellanic Clouds so that we might bask in the light of this glorious view. The best we can hope to accomplish is to make use of the information that reaches the earth from the heavens, carried on the backs of light, radio, and infrared waves, data that, when deciphered, allow a picture of the Milky Way galaxy to be constructed.

The concept that the Milky Way is a galaxy and might look something like other galaxies is a modern one, although a number of astronomers in the nineteenth century, such as William Herschel, had already suspected that the appearance of starry heavens indicated that we must live in a flattened disk-shaped "universe". Back in Galileo's era (the early seventeenth century) the concept of an "island universe" was unheard of, and even in the mid-nineteenth century it was but poorly developed.

Our terrestrial vision of the heavens is clouded by fog that drifts between the stars, a fog of interstellar dust that prevents light from the outskirts of the Milky Way from reaching the earth. Photographs of other spiral galaxies, such as M51 (Figure 17.1) reveal that the major dust regions lie along the inside edge of spiral arms defined by hot, bright stars, whereas photographs of edge-on galaxies (Figure 5.3) show the dust to be confined to this plane. The simplest explanation for both phenomena is that the dust is confined to the plane of the disk defining the galaxy. This creates the zone of avoidance (discussed in chapter 8).

Optical telescopes can peer along the disk of the Galaxy only as far as the nearest dust clouds, sometimes a thousand parsecs away, usually less. In a very few directions the Milky Way is transparent to light from greater distances, as in the constellation Carina, where stars as far away as 10 kpc have been seen. On the other hand, when pointed above and below the Milky Way telescopes can see outward to the edge of the universe for there is little intervening dust to obscure the view. The reason is that

FIGURE 17.1. The "whirlpool" spiral galaxy M51, in the constellation Canes Venatici. The dramatic alignment of the dark dust lanes along the inside of the luminous spiral arms, consisting of hundreds of millions of stars, is taken to be evidence for the propagation of a wave of density constantly orbiting the center of M51. This wave piles up matter which concentrates the dust into narrow lanes on the inside edge of spiral arms, as is evident, and within those dust lanes new stars are forming. M51 is a member of an interacting pair of galaxies and is about 12 million parsecs away. (National Optical Astronomy Observatories)

the Galaxy is an extremely flat disk, only about 100 to 200 pc thick. While the study of stars well away from the Milky Way—the high galactic latitude stars, which are unobscured by dust—has led to much knowledge of the properties of stars, very little about galactic structure on the largest scale has been derived from the study of starlight.

Most of the information crucial to forming a picture of the Galaxy comes from radio studies concerning the distribution of interstellar matter. Radio waves from clouds of interstellar hydrogen (chapter 16) allow the scale and structure of the Milky Way to be defined, while observations of molecular clouds (chapter 20) and infrared emission from warm dust highlight the structure as far away as the galactic center. Since these long radio waves from atoms and molecules pass unaffected through interstellar dust, they provide us with a radio "window" through which to see the entire depth of space in the Milky Way.

The Milky Way mapping adventure began inadvertently in 1932, when Karl Jansky at Bell Labs found strange radio hisses from space during his experiments to discover why transatlantic radio telephones suffered interference. This singular event marked the birth of radio astronomy, an undramatic moment as far as most astronomers of the time were concerned. It was up to a radio amateur named Grote Reber to use a dish-shaped radio reflector he built in his backyard to confirm that the Milky Way was a source of radio waves. But this radiation was to reveal little about the large-scale structure of the Milky Way.

Then, in 1944 in occupied Holland, Henk van de Hulst suggested, that a spectral line produced by neutral (cold) hydrogen gas between the stars should be observable at a wavelength of 21.2 cm. The existence of this gas had long been suspected, but was invisible to optical telescopes. This suggestion led to the birth of a major branch of radio astronomy; the study of the structure of the Galaxy.

The Rotation of the Galaxy

As was discussed in chapter 16, HI gas emits a spectral line at a frequency of 1,420.405 MHz. However, if the gas moves with respect to us we will observe a shift in frequency due to the Doppler effect (Figure 16.7). For gas moving away the frequency becomes less (the wavelength becomes longer); if it moves toward us the frequency increases (the wavelength becomes shorter). This Doppler shift when combined with a knowledge of the rotation of the Galaxy, allows the distance of the gas to be obtained. Figure 17.2 illustrates the rotation of the galaxy and its influence on the Doppler shift that affects the observations of distant hydrogen gas.

Imagine sitting on a large, slowly rotating carousel. As you look about you all the horses are moving just as you are, that is, around the center. None of them move any farther away or closer to you during a rotation.

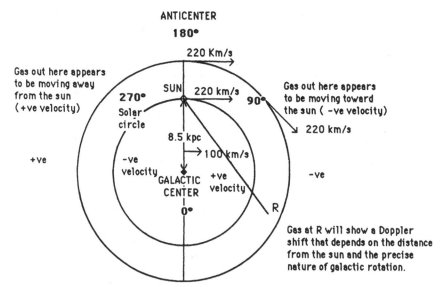

FIGURE 17.2. A schematic representation of galactic rotation and the way it influences the Doppler shift introduced to material in distant parts of the galaxy. Toward the galactic center and anticenter all material is moving parallel to the sun and shows no Doppler shift. In the neighborhood of the Sun gas is moving at 220 km/s around the galactic center located 8.5 kpc away. Closer to the galactic center the gas is moving less rapidly. Due to the nature of galactic rotation, material at great distances shows an apparent motion either toward or away from the Sun. By using an accurate model of galactic rotation the observed velocity of distant gas at any point, R, can be interpreted to give its distance. Numbers in boldface indicate galactic longitudes.

Imagine that the carousel takes one minute to rotate. Someone perched one meter from the center would travel a rather short distance, the circumference of a circle of radius one meter ($2\pi r = 6$ m.), so he is moving at 6 m/min. Now take up your position at the edge, say 10 meters from the center. You travel much further to get around once; in fact, you will be moving at about 60 m/min.

From any vantage point on the carousel, others sitting at different points appear to be travelling at different velocities depending on where they are located on the disk. Yet the distance between any two points does not change! This is what happens in the Milky Way and other galaxies such as M51 (Figure 17.1). The inner regions of the Galaxy, out to about 3 kpc, behave much like a solid wheel and exhibit *solid body rotation*. However, material further out does not travel as fast as is required to keep up with the inner regions, so that gas and stars lag behind matter closer to the center. This is known as *differential rotation*. The way the velocity of galactic rotation varies with distance from the center is known as its *rotation curve*. Such rotation curves can be determined for distant galaxies, where astronomers measure the velocity directly by observing the spectra of stars at different distances from the center of the parent galaxy and measuring their Doppler shifts. Our vantage point inside the Milky Way makes it tricky to determine the galactic rotation curve. An observer in M31 would have an easier time of it! By observing certain classes of stars at known distances (such as Cepheid variables) and measuring their velocities, the rotation curve of the Milky Way has been obtained.

The discovery that the galaxy is a rotating system harkens back to the 1920s. For some time before that a class of stars known as high-velocity stars were studied. They showed velocities not expected if they were gravitationally confined in Kapteyn's universe (a popular model at the time), defined by the stars visible around us. These stars had a predominant Doppler shifts which indicated velocities toward the Sun. The effect had been noted in 1781 by a Swedish astronomer, H. Gyldén, who noticed an apparent drift in stellar motions on one side of the sky while there was zero drift at right angles to this.

If the stars are very distant, then Figure 17.2 allows us to understand why they would show a preponderance of negative velocities in one part of the sky. At the time, though, stellar distances were poorly known and the concept of a galaxy barely nurtured. The recognition that high-velocity stars were distant and exhibited galactic rotation effects was precipitated by Jan Oort in Holland. In order for him to come to terms with the notion he had to consider that the Galaxy was far larger than Kapteyn's view had suggested. According to Kapteyn the sun was near the center of a relatively small star cloud. Shapley's model of the Milky Way as a galaxy, mentioned in chapter 16, made a lot more sense to Oort because it enabled him to recognize the consequences of galactic rotation in the stellar data. The theory of galactic rotation had been worked out by the Swedish theorist, Bertil Lindblad and observationally confirmed by Oort in 1928. He showed that two "simple" equations could be set up which defined galactic rotation in terms of observable properties related to stellar Doppler shifts (motions). The theory was later expanded to include HI observations, which could be used to derive the constants in the equation: the solar velocity around the galactic center and the distance of the sun from the galactic center.

In one fell swoop the size of the "universe" had been changed, the Sun placed firmly at the outskirts of a galaxy, now much larger than had previously been imagined, and the size of the Milky Way estimated.

By 1927 Oort had the distance to the galactic center to be 6.3 kpc and the center located in Sagittarius. The results of a huge body of modern research indicate that the sun is moving at 220 km/s around the galactic center 8.5 kpc distant. This distance has varied with time—not because of nature, but because of human inclination! The task of finding the distance to the galactic center is a considerable challenge, and in the 1950s and early 1960s was believed to be about 8 kpc. In 1964, as a result of new estimates of this distance, the International Astronomical Union (IAU) standardized the value at 10 kpc, which also made for ready reckoning with a simple number like 10. The velocity of the sun about the galactic center was taken to be 250 km/s. These values are important because they determine how we interpret observations of stars and gas in order to obtain distances. In subsequent years evidence grew that the standard values were not ideal and in 1985 the IAU gave its stamp of approval for revised values. The new, official distance of the galactic center is 8.5 kpc (which makes rapid mental arithmetic more difficult!), while the sun travels at 220 km/s in its circular orbit. Recently a suspicion that the solar distance is only 6.5 kpc has again begun to grow. Watch this space!

Thinking in 4-D

A radio telescope pointed along the Milky Way might observe a number of distant spiral arms each of which would produce a peak in the 21-cm spectrum, such as the one shown in Figure .16.2. This profile consisted of four distinct peaks at different velocities, which may be converted into distances using a model such as shown in Figure 17.2. Within three years of detecting the HI line in 1951 the Dutch radio astronomers had used this technique to interpret hundreds of HI profiles, and so produced the first map of the large-scale structure of the Galaxy. Today thousands of higher-quality profiles have been used to refine the work, and we will summarize that work here and produce a unique composite map that shows what the Milky Way looks like from afar.

The first map revealed spiral structure which proved that the Galaxy was a spiral much like M51 (Figure 17.1) and that the spiral arms wound their way out from the galactic center located in the direction of Sagittarius.

Data Presentation

The method whereby radio astronomers really "see" interstellar HI in order to determine spiral structure is a well-kept secret, perhaps because it requires thinking in three or four dimensions, preferably simultaneously, in order to sense what is going on. To aid in such mind contortions the data are presented in unusual, yet logical, ways: in the form of contour maps which look like geographical maps. An example was given in Figure 16.8.

A contour map may be used to show the altitude of a terrain, a population distribution, or a crop yield. A radio contour map may similarly be used to plot a number of parameters. For example, the contour lines may represent levels of the HI emission brightness at a given velocity, or the total amount of HI, or the density of the HI giving rise to the emission. Such maps reveal the "geography" of the neutral hydrogen (or of the emission from interstellar water vapor, formaldehyde, carbon monoxide, or other interstellar molecules) projected on the sky. These maps may then be compared with photographs to find, for example, whether dust clouds and HI "clouds" are coincident.

The presence of a hill or mountain on a contour map is taken to indicate the existence of a cloud of gas somewhere along the line-of-sight. The example of a contour map of the brightness of interstellar neutral hydrogen at a number of velocities in a certain area of sky shown in Figure 16.8 revealed many clouds as sets of closed contours. The diagram could equally well have been made as a photograph which displayed a high value contour as a brighter spot.

The challenge of mapping galactic structure involves using hundreds of contour maps, or the raw data that form such maps, with an additional twist in the way the data are considered. Maps with velocity replacing one of the position axes, thus showing the way the brightness of the HI emission varies with position and velocity, are used. How this is done is shown schematically in Figure 17.3. One coordinate may be galactic longitude (the measure of position along the galactic equator) and the other its velocity, a measure of distance (or depth). The radio telescope is systematically pointed at a sequence of directions along the Milky Way and at each position a spectrum is obtained, as the example shows. The spectrum at an adjacent position might have a slightly different appearance. These spectra are stacked together, each a cross-cut—a side elevation as it were—and a contour map then indicates what this stack would look like from above—the plan view. Long ridges in such maps are believed to indicate distant spiral arms; small sets of closed contours are clouds within those arms. Using a velocity—latitude contour map it is possible to display an enormous amount of information that is impossible to examine in any other way and still retain a sense of what is being observed. An example of a real contour map of this sort is shown in Figure 17.4, which was produced over a 13 hour observation with the 300-foot radio telescope at the National Radio Astronomy Observatory (NRAO) as it scanned along a line at fixed galactic latitude 5° above the central plane of the Milky Way.

Reference to Figure 17.2 allows one to recognize that the negative velocity gas over much of the map in Figure 17.4 must be due to hydrogen beyond the solar circle. The vertical ridge at zero velocity is due to hydrogen in the solar neighborhood independent of direction. Other lines

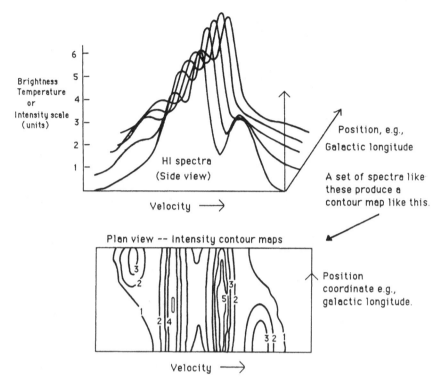

FIGURE 17.3. Making HI contour maps. A series of spectra such as shown in figure 16.10 (noise-free in this example) are obtained at a number of points along a line of, say, constant galactic latitude. These spectra are stacked side-by-side in a computer and then contour lines are found which represent levels of equal intensity, which are then drawn on a plan view where the two axes are one of position (galactic longitude in our example) and velocity (that is, the Doppler shift). Since velocity can often be interpreted as distance, these maps, called l-v plots, reveal the distance to the gas in question. The contours are in units of temperature. In this example four HI emission peaks can be seen. Two of them which cross the map are created by spiral arms, while the other two features are due to an HI cloud which may be parts of other spiral arms. It would require more such maps at other galactic latitudes or covering more longitude to make the identification.

of ridges or strings of clouds (closed contours) define distant spiral arms. Because the Galaxy is not flat, as will be discussed below, the interpretation of the map is made more difficult. The corollary is that if the galaxy were flat we would not expect to see distant hydrogen at negative velocities around longitude 50° to 80° at this distance (5°) above the Milky Way. Using Figure 17.2, the reader may roughly estimate how far above

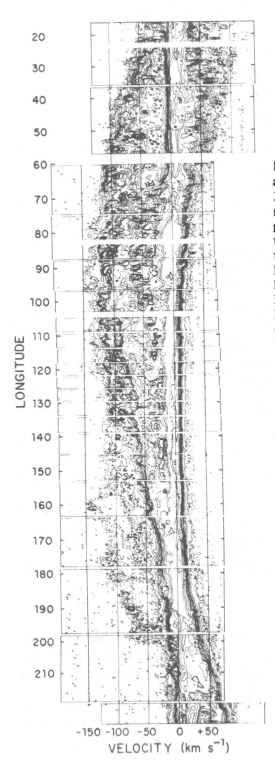

FIGURE 17.4. An example of an l-v plot obtained with the 300-foot radio telescope. The map covers a large range of galactic longitude and was obtained while holding the latitude fixed at 5°. It would be like scanning around you if kept your eyes 5° above the horizon. This map took 13 hours of telescope time to produce as the sky moved past the beam. A spectrum was obtained every 20 seconds, thus about 2,300 profiles were stacked side-by-side in the computer before the contour map could be produced. Galactic rotation causes distant matter to have negative velocities over much of this northern sky, which cover the two right-hand quadrants in Figure 17.2. The peak in the emission at zero velocity is due to emission from HI gas close to the sun which is seen all around us, while ridge lines may be identified in the other parts of the map, ridge lines which are taken to identify distant spiral arms. The resolution of this image as presented here is too small to allow the details to be seen, but when the map is blown up to be many feet across the existence of spiral arms becomes far more obvious. (National Radio Astronomy Observatory)

the galactic disk this hydrogen is located at the various longitudes. Note that the peaks at positive velocities at low longitudes are due to gas inside the solar circle.

Using a model for galaxy rotation allows the velocities of the peaks in the HI profiles in maps such as Figure 17.4 to be converted into distance. It must be stressed that the Doppler effect plays an important role because distant HI gas shows a Doppler shift depending on the distance of that gas (provided it is not situated toward or directly opposite the galactic center direction where all motion is across the line-of-sight). Thus, by knowing how the Galaxy rotates, the astronomer can interpret the observed velocities to derive distance and so create a map of galactic structure.

The Spiral Structure of the Milky Way

New 21-cm line surveys have revealed a startling image of the Galaxy. A plan view of a section of spiral structure based on 21-cm data obtained by northern-hemisphere radio telescopes is shown in Figure 17.5, but this reveals only part of the drama. The conventional view of the galaxy is that it is a flat, disk-shaped structure, about 12 to 14 kpc in radius and a few hundred parsecs thick. But the hydrogen in the Milky Way extends at least twice as far as the stellar disk, out to 20 kpc from the galactic center. Thus the Galaxy as defined in HI has a diameter of at least 40 kpc.

Maps of galactic HI structure are based on surveys of neutral hydrogen emission gathered over much of the sky. A radio telescope obtains a spectrum in a sequence of positions separated by 1°, say. To map an area extending 20° on either side of the plane of the Milky Way would include about 14,000 positions. In each direction the spectrum may contain 500 useful data points. Such a survey therefore requires the handling of at least 7,000,000 pieces of information. Several new, high sensitivity surveys are underway and they should lead to a much better understanding of the distribution of the interstellar HI on the largest scale. In the meantime, though, let us look more closely at all the recent maps of spiral structure that have been published.

As a special project for this book, efforts by about a dozen researchers to draw spiral structure maps (such as Figure 17.5) were collected together on a master map. The data were taken from reports based on the interpretation of 21-cm spectra as well as carbon monoxide data (discussed in the next chapter). In addition, clear evidence for spiral structure is obtained from studies of optical emission nebulae whose distances are obtained in a manner quite independent of velocity information.

Figure 17.6 shows the simplest spiral structure map that may be drawn consistent with about a dozen different reports on this structure. The

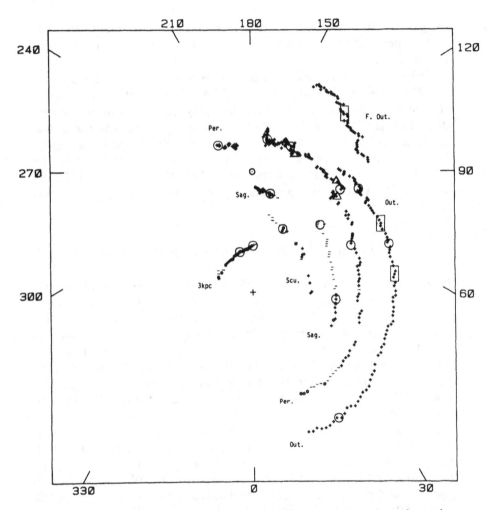

FIGURE 17.5. The velocities of ridge lines identified in maps such as shown in Figure 17.4 are translated into distances using a model for galactic rotation (see text). Spiral arms can then be drawn on a plan view of the Galaxy, as shown in this diagram, which used data obtained by a number of observers around the world. The astronomers who made this map, J.V. Feitzinger and J. Spicker of the University at Bochum, West Germany, noticed that in some areas it appeared as if the HI showed more complex structure than would be expected if the gas simply orbited the galactic center as part of the spiral density wave. They called these "velocity active regions" and indicated their presence with the rectangles and circles. A number of spiral arm segments have been labelled—see Figure 17.6 for details. (Jörg Spicker)

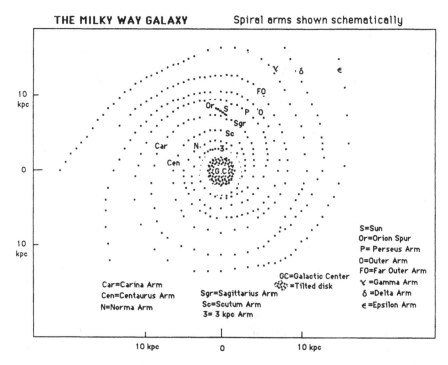

FIGURE 17.6. A composite spiral structure map produced by the author using about a dozen maps of the form shown in Figure 17.5, most of which were much less extensive. HI and CO spectral line data were used by the original researchers as well as information on the distances of HII regions. The inner circle represents the region where a tilted disk of HI and CO has been recognized. The four-arm spiral begins at what is known as the inner resonance ring, not indicated because it is more a theoretical construct than a ring of material. The Sun is located in the Orion spur between spiral arms at the coordinates 0, 2.2 in this diagram. Artistic license has made the spiral pattern fairly ordered, although the original data did show a very high degree of order. What is not well-known is the absolute distance of the most distant spiral features, many of which are not seen in the Milky Way, but reach either above or below it. Also, since the direction of the galactic center and anti-center allow no velocity and hence distance, discrimination, the maps made on the left- and right-hand sides do not overlap. Thus connections have to be inferred; the pattern presented here was obtained by making the simplest possible connections that allowed all the spiral features to be joined in an ordered manner. Nature may not, in fact, be so obliging.

Milky Way appears to be a four-armed spiral, in accord with suggestions made by several researchers, including A. P. Henderson of the University of Maryland. The arms in Figure 17.7 have been labelled according to various conventions that have appeared in the original papers. Notice that the sun is not located in a major spiral feature, but in a region of space called the Orion spur that stretches between spiral arms. From this oversimplified map we also get the impression that spiral arms are perfectly structured. In reality the gas and stars are likely to be distributed more irregularly, but the mapping done to date is not sophisticated enough to reveal the spurs and ridges that doubtless run between the arms, as is often apparent in other galaxies (see Figure 17.1). Also, joining the elongated features present in the data to form a four-arm spiral is controversial. All that can be said is that if one takes the well-observed features and connects them in the simplest possible manner, four arms are necessary.

The inner part of the Galaxy reveals an added touch. The gas there is distributed in a tilted disk inclined at about 20° to the the plane of matter beyond this region (Figure 17.6).

Comparison of the original maps showed surprising similarities, the greatest differences being in the distance of the outermost spiral arms. In defense of this attempt, it should be stressed that the map is qualitatively correct in that most of the features in one astronomer's map can be identified with those in another's. The ambiguity exists in the distances involved. These depend on the rotation curve of the galaxy, which is difficult to determine. Because there is, as yet, no unanimity on the form of the rotation curve virtually every observer has used a slightly different version. This only means that when the signature of a spiral arm is identified in the data of the type shown in Figure 17.4, the value for distance to the gas differs between researchers. Also, astronomers argue about whether ridge lines in the contour maps represent true spiral arms, but here we assume that they do. The recognition of a ridge line also depends on how sensitive the original observations were and thus not every observer has seen the weakest features. These points are made to stress the fact that the task of making a good spiral structure map is actually very complex. The largest telescopes operating with the best receivers have recently revealed features which went unrecognized in the early studies and which, in turn, has meant that the Milky Way as defined by HI has turned out to be larger than believed, as much as 60 kpc across perhaps.

The greatest challenge to constructing a composite map such as this lies is deciding how to join the segments of spiral arms mapped to the right or left of the galactic center, in whose direction no Doppler distance resolution is possible because everything is moving across the field of view. But again, the simplest connections were chosen and led to the four-armed spiral.

This exercise illustrates that it is possible to use the observations of interstellar matter to obtain a complete spiral structure map whose significance remains to be determined, as new data and more refined rotation curves allow distances to be more accurately set.

A Distorted Galaxy

The early 1954 HI surveys of the sky revealed an odd effect. Distant HI is not confined to the plane of the Milky Way, but reveals a warp or tilt in the outer regions of the Galaxy. The tilt is above the Milky Way in directions which would be to the upper right quadrant of Figure 17.2 (the northern skies) and slightly below the Milky Way in the upper left quadrant, the tilt reaching as high as 5 kpc in the north according to the most recent studies of the warp. It is much less in the southern skies.

But the initial shocks did not end, because further hints that all was not well with the Milky Way came in 1961, when Oort of Leiden began to search for infalling matter in the solar neighborhood. He hypothesized as follows. If star formation was still occurring in the galaxy, it was possible that matter was still being accreted by the Milky Way and fueling star formation. The existence of gas falling into the Milky Way might indicate that the process of galaxy formation was still taking place. To perform the search for this gas the radio telescope was pointed above the galactic plane and the receiver tuned to negative velocities to search for the HI coming toward the Sun. The desired signal was immediately found and at first was believed to have confirmed the prediction. However, the velocity of the gas was too large to be accounted for by this simple model. The HI appeared to be coming toward the sun at velocities as great as 180 km/s. Other gas, with velocities closer to 50 km/s, was also found, and it might conceivably be due to infall, but when the distribution of both types of 21-cm radiation across the sky was studied it appeared that the lower-velocity gas was most likely due to HI being pushed about by an old supernova shell. The high-velocity gas remained a mystery.

Further studies in both the northern and southern skies showed that these high-velocity clouds (HVCs), as they came to be called, were far more widespread than expected from an infall model. Clouds close to the sun move randomly with velocities up to 20 km/s and will therefore show Doppler shifts of this order, a limit set by the speed of sound in the interstellar medium. Even though interstellar space contains very little matter, at least compared to the air we breathe, it does have a sound speed defined as the speed at which particles can travel without creating a shock wave. Collisions between particles moving faster than the speed of sound would also ionize the hydrogen and make it invisible at 21-cm. Galactic rotation effects, on the other hand, can introduce Doppler shifts of as much as 150 km/s for distant gas in the Milky Way, as discussed with regard to Figure 17.2. But this gas should, in a flat galaxy, only be seen in the Milky Way and not as far above it as was observed by Oort and his collaborators. Yet the high-velocity clouds are seen to galactic latitudes of 60°.

Figure 17.7 is a schematic illustration of an all-sky map that shows where the high-velocity clouds are located. The key problem is that their

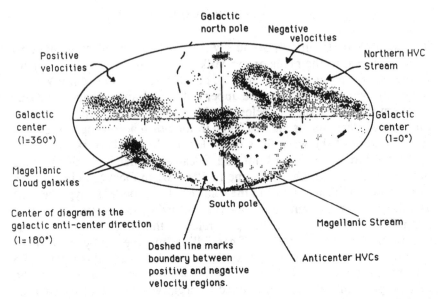

FIGURE 17.7. An artist's impression of a whole-sky map showing the location of the major high velocity clouds. The first to be discovered were part of the large Northern Stream that shows negative velocities in the upper right section of this diagram. The Magellanic Stream of gas sweeping away from the Magellanic Galaxies is seen in the lower region (see text). Positive velocity hydrogen moving more rapidly than the permitted 20 km/s for normal HI is seen in the left side of the diagram. Very high velocity clouds (up to −465 km/s) are seen between the end of the Magellanic Stream and the major high velocity cloud complexes.

distance is not known because no other objects, such as stars, nebulae, or other gases such as interstellar calcium, have been associated with them. In recent years the new surveys for high-velocity clouds revealed that some of the highest anomalous velocities (up to −465 km/s) are found closer to the Milky Way, in the southern skies. True infall should have produced the highest velocities near the poles. Also, gas with unexpectedly high positive velocities has been found in the southern skies. (Positive velocities indicate that the gas has a component of motion away from the sun.)

Where Are the HVCs?

The existence of the HVCs has been a problem since their discovery. It would have been solved long since if it were easy to estimate their distance, but this has proved to be an elusive challenge. In the meantime, a solution depends on dealing with the clouds in terms of what we believe we know about the Milky Way. If the Galaxy is flat, the high-velocity clouds, to be galactic, must be close to the Sun. But then, if the gas were close to the sun, why does it show such high velocities? One suggestion has been to imagine that the gas has been ejected for the disk of the Galaxy and is again falling back in, much like a fountain effect, but it fails to account for the distribution of this gas over the whole sky.

If the high-velocity gas is part of the warp in the outer galaxy, the warp has to be enormous to account for HVCs up to 50° above the Milky Way. What, then, would be the meaning of the warp concept? Or is the gas extragalactic, in the form of companion clouds to the Milky Way? If this is the case, why are the clouds so widely distributed across the heavens?

A few years after the first high-velocity clouds were discovered observations in various parts of the world revealed clouds moving at −100 km/s near the south galactic pole. At first it was believed that this gas had to be local material falling into the Galaxy. Otherwise, why would we expect to see it near the galactic pole? But no!

The story of discovery was to take a took a further dramatic turn in 1974 when observations made by Don Mathewson and colleagues in Australia revealed that the south-galactic-pole HVCs were part of an enormous stream of gas strung over much of the southern sky (see Figure 17.7). The "Magellanic Stream," as it came to be called, is 150° long and originates in the Magellanic Clouds, our nearest galactic neighbors located 60 kpc away. What had been previously believed to be empty intergalactic space was suddenly populated by a 100-kpc-long streamer of neutral hydrogen as much as 5 to 10 kpc wide in places.

It did not take the theoreticians long to account for the Magellanic Stream as the result of a tidal interaction between the passage of the Magellanic Clouds past the Milky Way. Gas was torn from these small galaxies during their encounter and the remains of the upheaval form a long stream of HI lacing through nearby intergalactic space. This same tidal encounter between the companion galaxies and the Milky Way may have created the warp in the Galaxy's outer edge.

Because the Magellanic Stream is clearly associated with the Magellanic galaxies, the HVCs at the south pole were unequivocally explained, but not as local infalling gas, but material from 50 to 60 kpc distant. Now we must ask whether the other HVCs, especially a huge stream seen in the northern skies (across the top right of Figure 17.7), may be a manifestation of a similar phenomenon, a vast intergalactic tail to our galaxy.

Continuing in the spirit of this chapter, a picture of the three-dimensional distribution of the spiral arms of the Galaxy was combined with the location of the two Magellanic Clouds and Stream, whose distances are known. A warp was added to the Galaxy and the combined data rotated in the author's personal computer.[2] Figure 17.8 shows the results of this exercise. The top-left diagram is the plan view of the Milky Way,

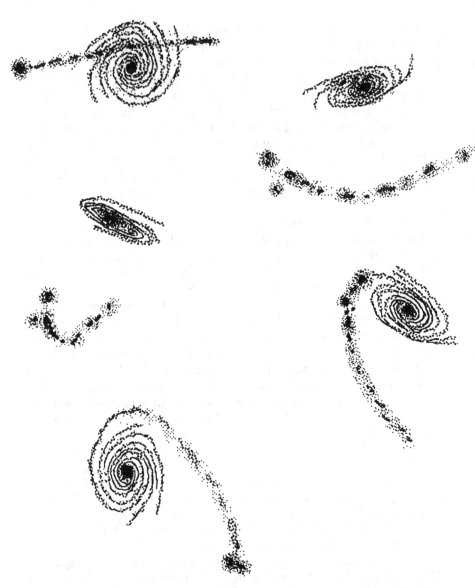

FIGURE 17.8. A collage of images relating the spiral structure of the Milky Way to the location of the Magellanic Galaxies and the Magellanic Stream in space. Top left is a plan view. The other four diagrams are perspectives obtained by rotating the point of view. An extraterrestrial in another galaxy would observe such views if it used a radio telescope of high resolution to map the HI. The two views in the lower part of the diagram may lead that ET to believe that the Magellanic Stream and Magellanic Galaxies were directly connected to the Milky Way.

touched up a little for effect, below which the Magellanic Stream sweeps across the sky. On the left, the two Magellanic Clouds are seen as a blend from this viewpoint.

Moving one's viewpoint in space involved rotating the configuration to simulate the view one might obtain from different directions, and this led to the other diagrams in Figure 17.8. The most intriguing of these are the lower two. These show the Magellanic Stream to be growing out of the most distant spiral arms in the Milky Way, a view that must provide an interesting challenge to astronomers in a distant galaxy that happens to suffer this orientation. Bear in mind that these diagrams represent the distribution of interstellar HI, although whether the label interstellar applies to the Magellanic Stream or not is in question, since no stars have yet been observed there (or in the most outermost parts of the Galaxy, either). This gas is, strictly speaking, intergalactic.

The enigma of the outer galaxy does not stop here. The outer regions of the Milky Way also grow thicker with distance from the center. Near the sun the Milky Way is about 200 pc thick, but it soon begins to swell until the HI layer is at least 5 kpc at the galactic boundaries (in the northern skies) although the boundary remains poorly defined. Furthermore, surveys aimed at finding more HVCs recently revealed gas travelling at velocities of up to -465 km/s in an area beyond the tip of the Magellanic Stream (Figure 17.7). R.J. Cohen, of Jodrell Bank, has suggested that these "very-high-velocity clouds" (VHVCs) are patches of gas torn from the tip of the Magellanic Stream, attached to neither the Galaxy nor the Stream. This model is shown in Figure 17.9 and includes more information than used to make the views in Figure 17.8. The thickening of the disk as well as its warp have been indicated.

By now the reader may feel very confused, which is just as it should be. This is the way astronomers dealing with the HVC mystery have been feeling for years!

To sum up, the Galaxy is warped in its outer regions, a hang-over from a close encounter with the Magellanic Clouds about 200 million years ago. An enormous streamer of gas now trails behind these companion galaxies as they continue to streak through space locked in orbit about the Milky Way. The Galaxy itself grows thicker, albeit more diffuse, at its outskirts and warps like a hat brim, up on one side, down on the other. So where do the high-velocity clouds really fit in?

In the early 1970s R.D. Davies at Jodrell Bank and the author, working at the NRAO, independently suggested that the Galaxy is so highly warped in the outskirts that distant spiral-like streamers rise way above the Milky Way and then turn down again as they stretch into space. This would simultaneously explain the location on the sky of the HVCs and their large negative velocities, a result of normal galactic rotation effects stamped upon very distant gas.

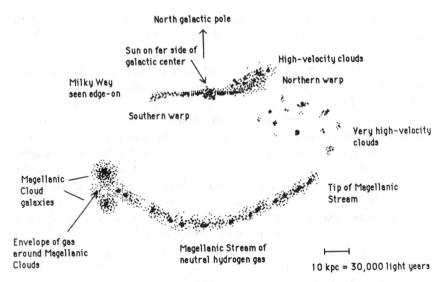

FIGURE 17.9. A view of the Milky Way for the far side of the galactic center looking back. seen on a pane that cuts The Magellanic Stream swings through space below the Galaxy. The spiral arms at the outskirts of the Galaxy flair to large distance above the disk and also grow much wider. The very high velocity clouds may be fragments of material in space beyond the end of the Magellanic Stream.

The Davies/Verschuur model was not quickly accepted, because it was impossible to reduce it to quantitative terms. The key issue is that the cloud distances remain unknown. Scientists, above all, want quantitative data upon which to build physical models. However, this author believes that the northern high-velocity clouds may represent a vast structure that begins as part of "normal" spiral structure and then flails from the edge of the Milky Way into "empty" intergalactic space. But, then, where does the Galaxy end? Does it even have a boundary we can define?

In the anti-center the HVCs are seen well below the disk. Perhaps that gas is so distant that it can hardly be considered part of an orderly warp anymore and must be pictured as a twisted stream of gas stretching away from the Galaxy, the analog of the Magellanic Stream, but in the opposite side of the sky.

The large-scale model for the HVCs received support from a survey guided by W.B. Burton, at Leiden, who used the 140-foot diameter radio telescope of the NRAO. He combined his data with southern sky observations made by Frank. H. Kerr and his collaborators at the University of Maryland to make a movie of galactic HI. The movie scans one's viewpoint around the sky and encompasses all the gas between +20 and −20° galactic latitude. The film shows the distant spiral arms warping above the Milky Way and then the most distant gas becomes very diffuse as it reaches high above the plane. The HVCs pop in and out of the scene at the Galaxy's edge in a manner consistent with the model of Verschuur and Davies.

To provide more variety to this fascinating mystery, Mathewson has suggested that the the Magellanic Clouds are not inexorably orbiting around our galaxy, but are currently visiting after a long journey over intergalactic distances. According to him they were in the vicinity M31 in Andromeda some 3×10^9 years ago. The Large and Small Magellanic Clouds have, in turn, had close encounters with each other and the Stream is indicative of material that has been torn out of the region of space between the two Clouds and left to trail behind them in their long intergalactic journey. In this spirit, he also suggested that the northern HVCs may be the remnants of a collision between our galaxy and another galaxy some 6×10^9 years ago.

We may well be living inside an interacting system, a spiral galaxy endlessly linked in a dance with the nearby Magellanic Clouds which, in turn, may have arrived from distant M31 (650 kpc away). Clearly great streamers of interstellar material reach into extragalactic space beyond the Galaxy's stellar frontiers. Yet the problem of the HVCs remains with us, in the quantitative sense, because until their precise distance are determined they will always remain a mystery to those who prefer not to depend on models for their description of reality.

Notes

[1] Sections of this chapter have appeared in *Astronomy Magazine* (January 1998, p. 26) under the title "Is the Milky Way an interacting galaxy?" and are reproduced by permission of *Astronomy Magazine*.

[2] The image was processed in a Macintosh Plus computer and the data rotated using MacSpin (D^2 Software), a powerful tool for such manipulation.

CHAPTER 18

Interstellar Magnetic Fields

A Great Physicist and a Classical Experiment

In 1897, in a physics laboratory at the University of Leiden, the sort of room where one imagines all major advances in physics in the nineteenth century to have been made—dark, gloomy, filled with equipment, wires, meters, and gauges—a young man of thirty-two performed an historic experiment. In years to come it was to have a profound bearing on astrophysical research. It would touch my life sixty-three years later when it set me on a path of discovery during which I visited the depths of disappointment and the heights of ecstasy.

Pieter Zeeman was a cautious physicist, a careful man, who in that year had not yet made a name for himself. Five years later he won the Nobel Prize for physics. On this day he would again try an experiment he had attempted unsuccessfully on several previous occasions. He wanted to know if the nature of the spectral lines produced by heated common salt changed when it was placed in a strong magnetic field. Common salt, when heated to incandescence produces a vapor that glows with a rich yellowish color due to two bright spectral lines, the sodium D-lines. The Dutch physicist believed something would happen but did not know what. Even though his knowledge of the way spectral lines were produced was current, it was founded on rudimentary theories which years later were shown to be incorrect. But who can know this in advance? The correct explanation for the formation of spectral lines required the development of atomic theory which was, as yet, a dream undreamt.

Zeeman had a hunch, a hope perhaps, that the magnetic field would produce an effect on the spectral lines. When he began his new experiment he felt motivated because he had just read a biographical sketch of Michael Faraday written by James Clerk Maxwell, both of them famous and respected physicists. Zeeman had been buoyed to discover that Faraday believed in a relationship between magnetism and light. This implied that magnetic fields might affect light, although Faraday also could not say how. Nevertheless the statement inspired Zeeman, who now set up his equipment even more carefully and tried again, this time proceeding with a sense of relief that came from the knowledge that he was not entirely alone in his views about magnetism and light.

He placed the flame of the Bunsen burner between the poles of a large electromagnet and suspended a piece of asbestos impregnated with common salt in the flame. The light from the flaring yellow flame was seen against a black background and was sent into a spectrometer. He peered into the eyepiece, examined the sodium D-lines, and moved the switch. Electricity ran into the giant coils, magnetic fields leaped into the air and penetrated the flame. The lines changed their appearance!

His spirit soared when the spectral lines widened to about three or four times their original width. He was ecstatic. Here was proof that light and magnetism were related. But what did it really mean? Now he worried whether he knew enough about his experimental apparatus to be sure that some spurious effect was not creating a subtle illusion, a wrong result.

Zeeman then proceeded to demonstrate how a great piece of classical science was done. He looked at the experiment from all angles to make sure it was behaving as he expected. His equipment was simple and elegant, certainly by modern standards.[1]

After his initial excitement had died down, Zeeman expressed suspicion that perhaps the field had changed the temperature or density in the flame to give rise to the effect. To test this, he placed the salt in a tube and raised it to a high temperature before switching on the field. The effect of the line broadening persisted. Then he created more elaborate tests to make sure that the effect was due to the presence of the magnetic field and not some other cause. All indications were that the effect was real.

When it came time to publish, the Dutchman presented an air of measured conservatism. "Possibly the observed phenomena will be regarded as nothing of any consequence,"[2] he wrote modestly. Zeeman's inner fears were expressed in this cautionary sentence. He was the pioneer, this was his discovery, and he had better make certain it was real.

We may also wonder whether he was simply protecting himself against being laughed at if the effect should prove to be of no significance, something we saw in Barnard's struggle with the notion of interstellar matter; or did Zeeman secretly fear that an unrecognized error of judgment or a subtle distortion of experimental technique had led to a spurious result?

The report he finally wrote is a beautiful example of a classic piece of research. In 1902 he shared the Nobel Prize for Physics (with H.A. Lorenz) for this work.

Theory and Observations

The reason that initially drove Zeeman to performing the experiment was based on an incorrect idea that there existed an "ether," an intangible substance believed to be the carrier of magnetic fields and light. Zeeman explained it as follows:

If the hypothesis is true that in a magnetic field a rotary motion of the ether is going on, the axis of rotation being in the direction of the magnetic forces, and if the radiation of light may be imagined as caused by the motion of the atoms, relative to the center of mass of the molecule, revolving in all kinds of orbits, suppose for simplicity, circles; then the period, or what comes to the same, the time of describing the circumference of these circles, will be determined by the forces acting between the atoms, and then deviations of the period to both sides will occur through the influence of the perturbing forces between ether and atoms. The sign of the deviation, of course, will be determined by the direction of motion, as seen along the lines of force. The deviation will be the greater the nearer the plane of the circle approximates to a position perpendicular to the lines of force.[3]

This description seems tortured to us, but it was all he had to work with. The theory of the ether would become increasingly cumbersome; in fact, early in the twentieth century the ether was shown to be non-existent (in the classic experiment on the speed of light by A.A. Michelson and E.E. Morley in 1904). However, Zeeman was using the best theory available to him and his work was a great example of how the true experimenter will never let theory stand in the way of the search. Zeeman's expectation that fields and light were related had borne fruit and later helped refine a great deal of theoretical work on both light and magnetism.

He then attempted to explain the line-broadening he had actually observed this time using Lorentz's theory, which claimed that small, electrically charged particles are present in all bodies and that light vibrations produced by an object are the result of vibrations of these "ions," as they were called. According to Lorentz's model, the charge, configuration, and motion of the ions would completely determine the state of the ether, and these ions would sense the field and thus the vibrations would change. Hence the nature of the spectral lines produced in the presence of a magnetic field would change.

Lorentz's model also predicted that the edges of the spectral lines should be circularly polarized and that the amount of the widening should also allow the ratio between the charge and mass of particles involved to be determined. It should be borne in mind that this experiment was carried out before the atomic theory involving protons and electrons had been developed.

So Zeeman returned to his experiment. In order to see the circular polarization he had to devise a way to observe the light along the same direction as the magnetic lines of force. This required that he drill a hole in the magnet and observe the sodium light through this hole. He then successfully observed the predicted circular polarization and concluded that it was the field that affected the vibrational period of the light, thus changing its wavelength slightly, which in turn broadened the lines. Lorentz's predictions turned out to be correct, although the model he used did not stand the test of time. And so progress is made.

Later developments in the theory of electromagnetism were to provide a far more rigorous explanation for the Zeeman effect. Michelson was to summarize the nature of the phenomenon which came to recognized as far more complex than Zeeman's initial experiments had suggested. Each sodium line was split into a triplet, not just broadened. The two outer lines of the triplet were polarized parallel to the field, and the inner one perpendicular to the field. When viewed along the field the two outer ones appeared circularly polarized in opposite directions.[4]

Zeeman's discovery stimulated a flood of research on both the experimental verification and theoretical aspects of the phenomenon. The Zeeman effect, as it subsequently came to be known, was exhibited in the light of many other substances in the laboratory. Then, in 1908, George Ellery Hale, Barnard's friend and colleague at Yerkes Observatory, measured the Zeeman effect in the light from sunspots, where fields with a strength of thousands of gauss were found. This led to the recognition that pairs of sunspots act as the poles of a magnet. In 1946 Horace Babcock at Mt. Wilson Observatory used the Zeeman effect to detect a strong field in the star 78 Virginis. This was the first observation of a number of the so-called magnetic stars. Their fields range from 100 to 34,000 gauss.

What Produces the Zeeman Effect?

Lorentz's original theory of the phenomenon was summarized as follows:

... that every molecule is a source of light (and) contains a single movable ion, which can be displaced in all directions from its position of equilibrium and is always driven back to that position by the same force, proportional to the displacement.[5]

The field was supposed to influence the motion of these "ions" but the theory was not so simple and required modification as further observations forced change. The emission of spectral lines was later explained in terms of the atomic theory as being related to the radiation of energy when electrons changed the orbits in which they circled the nucleus of the atom. In the presence of a field, the possible energy levels which the atom can adopt is slightly altered, with several new ones allowed due to the interaction between the atom's magnetic properties and the external field.

A classical way for understanding the effect as it applies to a simple atom such as atomic hydrogen (chapter 17) is to imagine that the hyperfine splitting is modified, because some electrons are spinning in a direction that is clockwise with respect to the field lines and others are spinning counterclockwise with respect to the field. When the field is introduced, the spin of some of them is speeded up and others slowed down a little. This results in a small change in the hyperfine energy levels in the hydrogen atom (chapter 17) and manifests itself as a splitting of the spectral line into two new ones centered on the primary one. But even this concept is too simplistic and only serves to give a *feel* for what happens. The formal explanation requires quantum theory, which is well out of our scope here.

From the Laboratory to the Interstellar Magnetic Field

At about the time I was doing the Zeeman experiment in an undergraduate physics lab, a radio astronomer at the Radiophysics Division of the CSIRO in Australia, J. Paul Wild, realized that the presence of an interstellar magnetic field might affect the properties of the 21-cm spectral line. If a sensitive experiment could be designed, the interstellar field could be measured. Wild's idea was a far cry from Zeeman's work, because Wild was suggesting that it might be possible to measure a 5 to 10 microgauss[6] (μG) field in clouds of gas thousands of parsecs away. Zeeman had worked with multi-gauss fields in the lab, and other astronomers had measured fields of as much as thousands of gauss in stars. How could anyone measure fields of only a millionth of a gauss in the depths of space? This turned out to be a challenge that could not be resisted at several observatories. Work began immediately to prepare the necessary equipment, for this was a variation of a classical experiment that could be applied in a new science—radio astronomy.

In the United States the National Radio Astronomy Observatory had just been founded and outside users were welcome and encouraged to bring their own equipment. At the University of Pennsylvania, T.K. Menon thought he would like to get involved and realized he needed a special antenna, or feed horn as it is called, which would be sensitive to circular polarization. To be mounted on the only radio telescope the NRAO yet had, an 85-foot dish, it would cost $9,000. Menon had good connections with a funding source which had just this amount left at the end of the fiscal year. However, the usual process for handling grant money required that the university charge overheads which would cut into the money and make it impossible to build the feed! So a compromise was worked out. The grant would be paid in the form of a check to the university and immediately endorsed to the manufacturer by a responsible person. It was promised that no other paper work would be required and so Menon obtained the antenna. Such smooth operation is all too rare in academia today.

Thus it came to pass that my fate was sealed, but I was still a student in a foreign land, oblivious of the existence of the NRAO, radio astronomy, interstellar matter, or Paul Wild's idea. To paint this scene more graphically, the science textbook in my high school said that space between the stars was empty!

The "Zeeman feed," as it came to be called, was first used by Frank Drake in his 1961 search for radio signals from extraterrestrials (Project Ozma), a search that gained much notoriety. Menon's subsequent efforts to detect the Zeeman effect met with no success. The idea for detecting the field was unfortunately well ahead of available technology. Nor did similar experiments succeed at Jodrell Bank, where, in 1958, they had also begun to make plans to seek this version of the Holy Grail. It was to Jodrell Bank that my path would lead.

Magnetic Fields in Space

There were many reasons for wanting to know the strength of the interstellar field. In 1949 Enrico Fermi[7] suggested that the existence of such a magnetic field would explain why cosmic rays from deep space were striking Earth with great energy. These particles travel near the speed of light and in order to explain their high energies he had to invoke some way to inject more energy into them even as they sped through space; otherwise the particles would quickly slow down and never reach the earth. His suggestion required fields of 5 to 10 μG. By way of metaphor to illustrate Fermi's model, the fields were supposed to accelerate cosmic rays because they bounced between moving clouds of hydrogen gas in the way a ping-pong ball bounces ever more rapidly between two paddles as they are brought closer together.[8] In the ping-pong case the energy for speeding up the bouncing ball comes from the movement of the paddles. The energy for accelerating cosmic rays was to come from moving HI clouds. Fermi's voice of authority was enough to cause the idea that fields existed in space to fall upon ready ears. Such a field might also produce the aligning mechanism that was required to explain why starlight was polarized.

In 1948 John S. Hall at the U.S. Naval Observatory and W.A. Hiltner at the McDonald Observatory of the University of Texas had detected the polarization[9] of starlight as an unexpected sidelight to the measurements they were making. It had been predicted that the light from stars should be polarized in their atmospheres, with the polarization following the curvature of the star's edge. Observations of eclipsing binary stars would reveal the effect. When one star passed in front of the other the polarization in the atmospheric rim of the background star should become visible. Instead, Hall and Hiltner found that other stars, not their candidates, were polarized, and that the amount of polarization depended on reddening, a measure of how much dust lay between those stars and earth. It was soon suggested that the field interacts with elongated interstellar dust grains and pulls at them, so that they align along the field. Starlight passing through the dust then becomes polarized as it interacts with the aligned grains, and the angle of the polarization was believed to be related to the magnetic field direction in clouds of interstellar matter. Such data only gave information about the field's direction, not about its strength, a crucial parameter in the theoretical models.

The Quest Begins

Figure 18.1 shows a schematic diagram of the production of the Zeeman effect at 21-cm. Unlike the laboratory experiment, in which the line splitting can be seen, the splitting of the 21-cm line was predicted to be extremely tiny, only a few tens of hertz at a frequency of 1,420 million Hz. It is not possible to measure this small splitting in an absolute sense, because the line width is at least 10,000 Hz. However, since the components of the Zeeman split lines are circularly polarized in opposite senses (right- and left-hand), it is possible to subtract the signals received by each of the two polarizations and thus (hopefully) detect the tiny difference to be expected as illustrated in Figure 18.1.

In order to follow the checkered history of this experiment, the reader should be aware that radio astronomical signals are subject to "noise," as was discussed in chapter 16. From the start it was recognized that the predicted Zeeman effect would produce a signal that was below the noise in a single observation. It would take dozens or even hundreds of hours of observation on the same spectral line to offer any hope that the weak Zeeman signal would poke through and be seen.

In 1961 I arrived at Jodrell Bank and began my involvement in the quest for the interstellar field. We had no computers and in the first experiments only a single-channel receiver was used. All the data were recorded on paper charts. The frequency of the receiver was scanned by

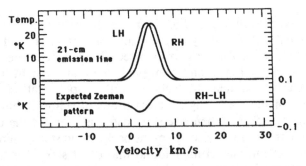

FIGURE 18.1. The principle of measuring the 21-cm Zeeman effect due to a magnetic field in a hydrogen cloud seen in emission. In the presence of a field the line is split into two components of opposite circular polarization. By switching the receiver between these two polarizations and subtracting the two signals it is possible to detect a faint radio signal which is directly proportional to the strength of the interstellar magnetic field in the cloud.

a small motor and the line profile and "difference spectrum" (Figure 18.1), which would reveal the Zeeman effect, were plotted on these charts. Then graduate students would rule baselines in by hand and read off the deflection from the record with respect to this baseline to the nearest millimeter. Hundreds of passes across the spectral line were made and all data on the chart records had to be read off. Thousands of numbers were averaged by hand. We became very adept at adding, dividing, and subtracting. At last all those years of undergraduate study—math, physics, applied math—were paying off!

In 1962 our laboratory, located 120 feet above ground in a support tower of on the 250-foot telescope (Figure 16.1), was supplied with a paper tape machine which spewed out a roll of tape in which punched holes represented the numerical value of the data. After the observing run was over we had to book time on the Atlas computer at the University of Manchester to read the tapes. Now we could automatically add and average the data, and this took many eight-hour night shifts at the computer.

Between 1959 and 1965, radio astronomers in at least four observatories became involved in the search, but the Zeeman effect in space was never detected. One false alarm unfortunately reached beyond the privacy of our laboratory. When the announcement of the spurious detection was made to the press a reporter called and asked whether it had been like sitting a bath tub and suddenly shouting "Eureka!" Together with another graduate student involved in this fiasco, I did not believe the detection, so the question was moot.

The problem in detecting extremely faint radio signals from space is that unrecognized and systematic effects sometimes lurk below the noise and rear their ugly heads only during very long integration times required to reduce the noise. (Today an added problem is that man-made interference can perform the same role of ruining the data.) Electronic equipment has a way of generating its own reality. Only extensive experience in the vagaries of radio telescopes and receivers allows the astronomer to perform effectively and so minimize the chances that systematic problems will confuse the results.

The 21-cm Zeeman effect measurements the world over were, from the start, ruined by spurious effects which took years to be identified. Some of the problems involved an imbalance in the polarizations in the antenna which could be dealt with by making adjustments while observing. This, at Jodrell Bank in the early days, required endless safaris from the laboratory to the focus of the big dish. First, a calibration observation had to be done. Then the telescope was driven to the zenith. The graduate student on shift took an elevator up a couple of levels, walked across the catwalk below the dish, climbed up into the bowl, took the rack and pinion cage on its four-minute vertical journey to the top of the 60-foot mast, and opened the "focus box". Here rested the critical amplifier which would cease to work if one looked at it. The student would then fiddle with a switch or antenna, travel down the mast in a four-minute journey, climb back to the lab, then request that the telescope again be pointed at the radio source and see if the fiddling had had a beneficial effect.

In 1962 the 21-cm Zeeman work at Jodrell Bank benefitted because we were able to gather six channels of data simultaneously. Because the use of the computer was not yet very convenient this meant that six times as many numbers had to be read off and added by hand!

Trauma Rewarded

When I left England for the NRAO in 1967, the 21-cm Zeeman experiment had not succeeded anywhere in the world, although by now a variety of observers had used thousands of hours of telescope time in the search. Their stories later surfaced and had a common theme—they had all seen noise and spurious effects, most of which confounded them.

"You will surely do the Zeeman experiment at the NRAO," colleagues said, after I arrived.

"No," I answered, recalling the trauma of years of struggle with no results and a Ph.D. thesis built upon this work. But the pressure continued and I could not ignore the fact that the 140-foot telescope of the NRAO was a beauty and it was about to be equipped with a 384 channel receiver (compared to the meager 6 channels we had used at Jodrell Bank). I also heard that a special feed for the Zeeman experiment was lying around somewhere. This was the one that Menon had had built nearly a decade before. It turned up in a warehouse where it had been mothballed. I learned someone else had attempted the experiment on the 140-foot and given up trying to understand the chaos in his data. His personal report to me was not encouraging. I nevertheless decided to test the feed and asked for two days on the telescope. "A good idea," one of the referees of the proposal said, "While you are at it, take ten days to do the experiment." This may have been the only time that an astronomer was granted more time than he requested to do an experiment, especially one that he was by now so reluctant to deal with because seven years of blank do not create the thrill of discovery which helps motivate one in research.

In May 1968 I began to observe with the new 384-channel receiver, but the data reduction programs had not yet been written. This posed an added challenge because one cannot observe effectively without having some idea of what is going on. In modern parlance, I had no on-line capability to look at the data, except of a most rudimentary sort. This consisted of a printer that provided columns of 384 numbers at the touch of a button. Someone touched that button every 30 minutes for the next ten days. My well-honed skill at adding numbers served me well. For weeks I did very little other than that: add, divide, and occasionally eat and sleep. Thousands upon thousands of numbers flowed before my eyes because I did not have the patience to wait for computer programs to be finished. In any case, no one should ever trust a new program. I needed to know what it would tell me before it did so.

A month later, on an airplane, I was still adding and dividing and subtracting, but by now there was a hint that a signal was present in the direction of the Crab nebula radio source. A dense HI cloud absorbed its 21-cm radiation and a field appeared to be present in the data on this spectral line. But the signal was very marginal, not much above noise, and less than a tenth as bright as the false alarm claimed in 1962. (This spectral line had always been regarded as the most likely candidate in which to find a Zeeman effect. The other was a deep absorption line produced by local HI gas in the direction of the Cassiopeia A radio source, the remnant of a supernova explosion that occurred 350 years ago.)

By July 3, all the programs had been written, tested, and debugged, and were finally ready for the grand analysis. Left to run overnight, the computer churned through ten days of data. The next day, I went in to check on the printout. The results could not be automatically graphed, so I plotted them by hand.

First I plotted the Crab results. They looked much as I had seen on the plane—reassuring, but not convincing. At least the computer knew how to add and subtract as well as I did. Then the data for the local (zero velocity) absorption line in the direction of Cas A was plotted. The difference spectrum was as flat as a pancake (see Figure 18.2); no spurious signals, no Zeeman effect. The field was less than one microgauss in those HI clouds. Too bad for Fermi and all theorists who required 10-μG fields to align dust grains.

The Thrill of Discovery

Before I went home for the afternoon (after all, it was a national holiday), I thought I would plot some more numbers. The 384-channel receiver had covered such a large frequency range that the absorption lines due to the 3-kpc distant Perseus spiral arm (chapter 17) in front of Cas A just happened to be included at the edge of the band (Figure 18.2). In my personal number crunching I had at first ignored these lines because they were not a likely place for a field, a conclusion based upon prejudice built upon habit. In any case, I had added so many thousands of numbers that I had left some work for the computer.

As I plotted the numbers on the Perseus arm feature in front of Cas A, I could not believe what I was seeing. There it appeared! A huge Zeeman effect! It may have been only an inch or so on the graph I was drawing, but this was the culmination of an eight-year quest! That was, to my eyes at least, an *enormous* field!

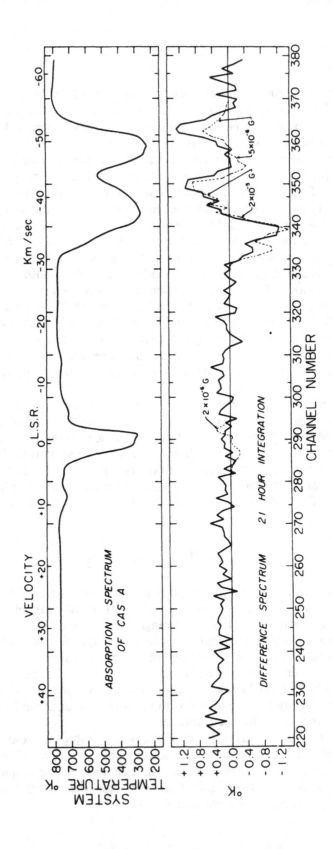

When I managed to drift down from the ceiling I calculated that the data revealed fields of 10 and 20 μG in two clouds. What would Pieter Zeeman have thought of that! These were signals due to Zeeman effects in diffuse clouds in very distant space. My workaholic office mate who was, as usual, at his post on that holiday, was quickly convinced that the detections were real as I babbled to him about the results.

Elation! Ecstasy!

The data on the other source, the Crab nebula, revealed a field of only 3.5 μG in a cloud 1 kpc distant. For comparison, the field measured by Babcock in some magnetic stars was 10 billion times as strong, yet both were revealed by detecting an effect that Zeeman had feared might "be regarded as nothing of any consequence."

And Again ...

Later that year I was on the telescope again, this time with an on-line capability that allowed me to see data in real time. Within half an hour of starting observations on Cassiopeia A the Zeeman signal appeared out of the noise. It had recently been confirmed at Jodrell Bank, so now I could use it to make sure the system was working correctly. The Zeeman effect on Cas A had now graduated to being the calibrator!

Then the telescope was pointed at the Orion nebula, whose radio image is shown in Figure 18.3, and in front of which lies a dense HI cloud. Within fifteen minutes of beginning the integration another field effect became obvious. Or was it? This signal was very strong and might well have been spurious. Even as I planned to make the necessary measurements to confirm that the detection was real, the thrill of discovery again coursed through my veins. This was an on-line detection, the first ever, after a half-dozen experiments spread over nine years during which no signals had ever been seen; only noise or spurious effects. The thrill is difficult to relate. This was far more exciting than the July 4 discovery because this was live action at the telescope!

The most ironic fact was that the signal was so strong that if we had looked for it at Jodrell Bank five years before we would probably have seen it, even with the crude system available at the time.

FIGURE 18.2. The actual data obtained in the first successful detection of an interstellar magnetic field in the direction of absorbing HI clouds in front of the Cassiopeia A radio source. The local cloud shows no evidence of a field, which is believed to be less than 1 microgauss (μG) here, while the clouds in the distant Perseus spiral arm reveal the presence of strong fields of up to 20 μG. These data were the result of a 21-hour integration required to reduce the noise levels (see chapter 16).

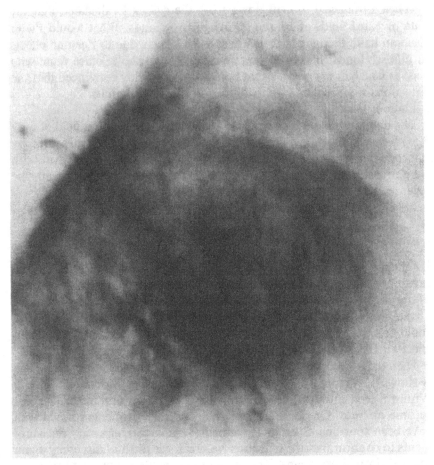

FIGURE 18.3. A "radiograph" of the Orion nebula (Figure 1.4). In this image the intensity of radio emission has been converted to a light signal and directed onto a photographic plate. The Orion nebula is a radio source by virtue of the thermal emission radiated by its hot gases (at 10,000 K). An invisible cloud in front of this nebula contains a field of 50 μG. (F. Yusef Zadeh)

And the Agony of . . .

To make sure that the signal was not spurious I had to make further tests. After adjustments to the feed were made I asked the telescope operator to re-observe Cas and Orion and left for dinner. When I returned and looked at the recordings made in my absence, they revealed *no signals* at all! I was stupefied beyond belief. Only a couple of hours before I had seen the signal on Cas A, which had been confirmed at Jodrell Bank, and then one on Orion. Now both were gone!

I sank into the chair by the telescope operator and poured out my concerns.. Was some deeply hidden spurious effect, capable of almighty nastiness, making its presence known? I told him about the checkered history of the experiment and how this latest twist made no sense at all. As I related the tale I happened to stare at the receiver rack and noticed that one switch among dozens was incorrectly set. After our adjustments to the antenna it had not been reset.

Excitedly I leaped up and set it right. "Back to Cas," I cried, knowing what I would see. Fifteen minutes later, yes, there it was!

"Now for Orion," I yelled across the control room. The telescope slewed to the source. This was to be the acid test for the new detection. If it came through again it had to be real. If it did not, the system was in trouble and my night would be ruined. If it was still there, well then . . .

I stood, my eyes impatiently glued to the cathode ray screen that showed the integration building up. After five minutes there was a hint of a signal. Time dragged and stretched out. A little green spot of light scanned the screen and drew slightly less noise each time. Ten minutes. A signal? Yes, yes, if you know what to expect. Twenty minutes—without a doubt! Half an hour. . . .

"Its there! Its real!" I shouted at the operator, who had joined me to stare at the screen, awaiting the signal from space. I danced for joy. The signal was roaring in!

Roaring? Well, not quite. In radio-astronomical parlance it was 1.2 K bright, but after an eight year drought this whisper carried a flood of information. The signal I was staring at was produced by a 50-μG field in a cloud in front of the Orion nebula and had travelled 1,200 years to get to the telescope.

Reprise

Eleven years had passed since Paul Wild's note on the possibility of doing the experiment. Somewhere between 6,000 and 10,000 hours (8 to 14 months) of telescope time, world-wide, had been used in the search, all without success.

A weak field effect was found in the direction of M17 (see Figure 19.1) and in due time all were confirmed either at Jodrell Bank or in Australia. During the next few years no more detections were made. Since then many more thousands of hours of dedicated time on a smaller radio telescope—operated by the University of California, Berkeley, under the guidance of Carl Heiles—have been applied to the task. In 1982, he and his collaborator, Tom Troland, reported fields of 7 to 10 μG in two HI clouds. In 1986, I returned to use the 140-foot with its vastly improved sensitivity, and at the time of this writing field detections have been reported in a total of ten directions, a meager harvest for the thirty years work, world-wide, since the experiment was first suggested.

Other Views of the Field

Today the properties of the interstellar field come not only from the 21-cm Zeeman effect work. Observation of the polarization of pulsar signals which are influenced on their way to earth by interstellar fields suggested to Richard Manchester in Australia that the field in regions of space permeated by ionized material may be about 3 μG. The Zeeman effect in the OH spectral lines has also been observed and has revealed fields as high as 100 μG, but these are in dense dust clouds associated with star formation.

Radio waves from the Milky Way, which were originally discovered by Karl Jansky in 1933 and led to the birth of radio astronomy,[10] are polarized, and study of this property gives information about the interstellar field direction. These measurements are also notoriously tricky (and hence unpopular as a research topic to which new students might dedicate years of work) and much remains to be done.

The properties of the interstellar magnetic field remain little understood, although the field may be of fundamental importance in determining the appearance of virtually everything in space, whether it be the dust or gas clouds, spiral arms, or even the shape of galaxies. If only the theory and observation of the fields were not so difficult, the field would be firmly incorporated into all models related to interstellar matter. The truth is that field strengths are poorly understood, and the influence of the fields in theoretical models almost always ignored.

Today the average interstellar field as revealed by the 21-cm Zeeman effect measurements or interpretation of pulsar data is too low to account for the alignment of dust grains to explain the polarization of starlight. Also, Fermi's need for a 5-μG field to accelerate cosmic rays in moving interstellar clouds has turned out to be without foundation. Cosmic rays are now known to be accelerated in shock waves driven into space by supernova explosions. Fermi might have been on the wrong track; nevertheless, his suggestion that fields exist in space stimulated a lot of work that led to greater knowledge of interstellar matters.

Notes

[1] Your author had to perform this experiment as a graduate students in physics. I had to leave my wristwatch outside the room lest it be damaged by the field of the magnet. I sat in the dark beside a large electromagnet, a Bunsen burner glaring yellow from salt held in its flame, while peering through a small telescope at the sodium D-lines. I switched on the field and it was exciting to see the sodium lines change their appearance. Little did I know that this same experiment, shifted to an astronomical context, would some day determine my life as an astronomer.

[2] Zeeman, P. (1897) "On the influence of magnetism on the nature of light emitted by a substance." *Astrophys. J.*, 5: 332.

[3] Ibid.

[4] Michelson, A.A. (1898) "Radiation in a magnetic field." *Astrophys. J.*, 7: 131.

[5] Lorentz, H.A. (1899) "Considerations concerning the influence of a magnetic field on the radiation of light." *Astrophys. J.*, 9: 37.

[6] A gauss is a unit of magnetic field strength. The earth's field is about 0.3 gauss

[7] Fermi, E. (1949) "On the origin of cosmic radiation." *Physical Review*, 75: 1169.

[8] This metaphor is borrowed from Martin Harwit's book, *Cosmic Discovery*, which brilliantly summarizes the progress of astronomy in highly readable fashion and makes cogent comments concerning the future of this science (Cambridge: MIT Press, 1984).

[9] Polarization refers to the favoring on the part of an electromagnetic wave to concentrating its energy in one plane rather than another. The simplest analog is that if one holds a rope at two ends and flips one end up and down a vertically polarized wave runs along the rope. Flipping it horizontally sends a horizontally polarized wave along the rope.

[10] Verschuur, G.L. (1987) *The invisible universe revealed—The story of radio astronomy,* New York: Springer-Verlag.

CHAPTER 19

Molecules and Interstellar Matter

Dark Markings and Interstellar Molecules

Now we move to very recent times, the last decade or so, when the study of interstellar matter began to reveal an event which until recently astronomers believed they might never see: star birth itself. Edward Barnard could never have imagined that the "holes in the heavens" might someday reveal so much about this process and that the research activities of hundreds of astronomers would be devoted to studying this topic.

Since 1970 the types of dark clouds discovered and cataloged by Barnard have produced the richest harvest of discovery ever gathered in astronomy. These clouds contain hidden protostars—objects that will form stars—as well as young stars that have just emerged from their nests. Stars are spawned in swirling masses of dust and gas in the dark clouds where the fetal stages before starbirth are hidden from prying eyes. But they are revealed to the insistent probing of radio telescopes and infrared detectors. It is within the dark clouds (Figure 19.1) that the birth of stars from dust, gas, and molecules proceeds. Here protostellar objects, ready for stardom, burst from their shrouds and shrug off layers of matter in energetic jets that stream outward and indicate to viewers on distant planets that a star is about to be born (see Figure 21.3).

The most stunning side to these dark clouds is that radio observations have revealed the characteristic fingerprints of dozens upon dozens of molecules, a large fraction of which are organic (carbon-based). In Barnard's day this would have sounded like a ridiculous idea, and even as recently as 1950 astronomers did not seriously expect that molecules of any degree of complexity (beyond a combination of two atoms) could exist in space. The discovery of these complex molecules between the stars has profound implications for our understanding of the origin of life, because the dark markings that Herschel and Barnard once thought were vacancies contain the same molecules required for the earliest stages of life.

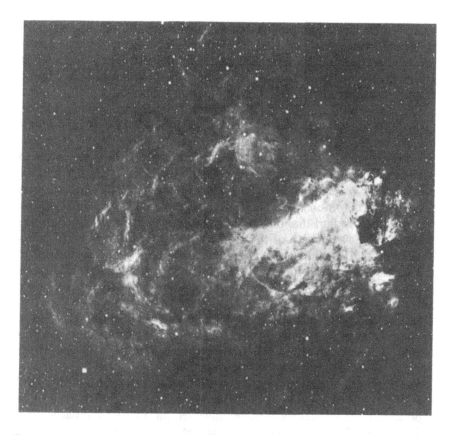

FIGURE 19.1. The Omega or Swan Nebula known as M17 (NGC 6618) in Sagittarius. A region of star formation laced with interstellar matter in the form of gas and dust. The nebula is 4 parsecs across and located 1,900 parsecs away. The filamentary structure is produced by the heating of interstellar matter to incandescence by the young stars that are the energy source in the nebula. The dust clouds are rich in interstellar molecules. The luminous material gives rise to an HII region which produces a strong radio source. Photo taken by the Kitt Peak 4-m telescope. (National Optical Astronomy Observatories)

The Discovery of Simple Molecules in Space

William Huggins and Angelo Secchi were the pioneers in the field of spectral analysis of the light from nebulae in the mid-nineteenth century. Secchi had no way to record spectra on film, so he had to describe in words and sketches what he saw. He created the first scheme for classifying stars according to the appearance of their spectra. Huggins was similarly hampered in having no photographic plates to work with when he observed the spectra of eight planetary nebulae. In endeavoring to identify the various spectral lines by making comparisons with those produced in the laboratory by known substances, he ran up against a strange phenomenon: A mystery line was present. The mystery line occurs in the green part of the visible spectrum and gives many nebulae their greenish glow, such as can be seen when the Orion nebula is viewed through a moderate-sized (e.g., a 16-inch) telescope.

The story again goes back to William Herschel, the venerable musician turned astronomer who seems to have had his fingers in every astronomical pie of his age. He had suggested that the "shining fluids" that constituted the bright nebulae consisted of elements known on earth, in particular in the atmosphere. Huggins therefore tried to associate his observations of the spectra from planetary nebulae with spectral lines produced by common atoms such as oxygen, hydrogen, and nitrogen, but was surprised to find that only one bright line could be identified in this manner. The wavelength of the mystery line was close to one of many lines expected from nitrogen, although none of the others were seen in the nebular spectrum. Nor were lines from other common terrestrial gases present. To explain the bright line he therefore proposed the existence of a new element, not found on earth, to be called "nebulium."

In due course several spectral lines due to hot hydrogen were found in the spectra of the planetary nebulae, which hinted that Herschel was correct. Then Secchi found nebulium in emission nebulae as well. However, he never lived to see the line identified.

Years later the nebulium line was attributed to a so-called "forbidden" line of doubly ionized oxygen—forbidden because it cannot readily be produced in the laboratory. Nature, however, has its own standards. Forbidden lines of other elements were subsequently recognized.

The field of optical spectroscopy blossomed. In the early part of this century the spectra of stars and nebulae were studied in detail and led to a storehouse of information on the chemical make-up of stars and nebulae. But although the presence of interstellar calcium had been known since 1904, very little further thought was given to spectral observations of stuff between the stars.

Then, between 1937 and 1941, the optical spectral lines from three interstellar diatomic molecules (CH, CN, and CH^+, which is the ionized form of CH) were discovered. As early as 1934 several broad spectral patterns, known as diffuse bands, due to interstellar matter were seen. Their precise origin remains a mystery to this day although they are very likely due to a variety of PAHs. Thirty-nine bands lie between wavelengths of 4,400 and 6,850 Angstroms (the optical unit of wavelength).

The mystery of the diffuse bands has caused most astronomers interested in the subject to tread warily. However, certain others have ventured bravely into the unknown in search of this Holy Grail. Their quests have led to an almost bizarre range of suggestions to account for the bands. Porphyrins, chlorophyll, carotene, and even bacteria have been pronounced guilty. Such notions have fallen on deaf ears and left the majority of astronomers vastly unconvinced. This does not imply that the suggestions may not yet turn out to have some merit!

Laboratory work has shown that sulfur atoms cause patterns of a bandlike nature, and these are widely suspected to be present in space. While at least four different carriers, such as sulfur, are probably involved in producing the diffuse bands, many astronomers believe that the problem may never be totally resolved, because the laboratory simulation work required to prove an identification is impossibly difficult. Part of the problem is that a wide range of complex molecules is capable of producing similar wide bands.

By 1950, in addition to interstellar calcium and sodium (which were found much earlier in the century), lines due to interstellar atomic potassium, iron, and titanium had been observed in space. It was at last clear that space was far from empty; it contained dust, certain atoms, and a few simple molecules. Much of space is filled with matter far more diffuse than the material that produces Barnard's dark markings, as is now dramatically revealed by IRAS (Figure 19.2).

Few astronomers gave serious consideration to the possibility that more complex molecules existed out there. As late as 1951 several experts concluded that the molecular formation rate would be so slow and destruction by ultraviolet so rapid that it was, in fact, impossible for interstellar molecules to exist. Even when important lists of molecular spectral lines in the radio astronomer's wavebands were published by Charles Townes in the United States and I. Shklovsky in the Soviet Union in 1951, no one took much notice.

The Modern Era

Today the situation has undergone a total transformation. Nearly eighty different species of molecules (listed in Table 19.1) have been identified in the dark interstellar clouds or in envelopes of matter around cool, giant stars. The length of this list is determined only by technical and theoretical limitations, not by the variety extant in space. Radio spectroscopy has also revealed hundreds of unidentified spectral lines suspected of indicating the existence of more complex molecules than those whose spectral lines have been studied in the laboratory up to now.

The observations of interstellar molecules by radio astronomical or infrared detection follows the same principles of radio spectroscopy described in chapter 17. Just as is the case for HI, molecules have characteristic signatures that can be identified in the laboratory and then sought for in space with the aid of telescopes. Radio spectroscopy, in turn, is analogous to optical spectroscopy which allows the chemical composition of stars to be studied and which originally revealed the existence of interstellar calcium and sodium.

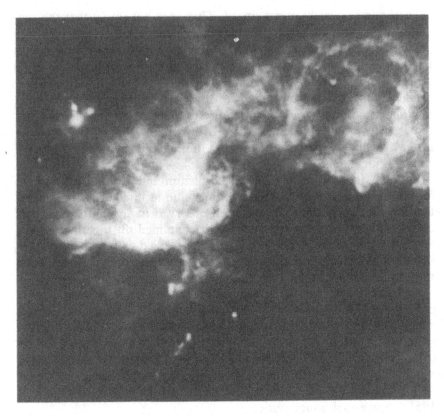

FIGURE 19.2. Interstellar cirrus revealed by the IRAS satellite. The circular filamentary structure is almost certainly related to a very old expanding remnant of
a supernova. The sky coordinates for this image range form galactic longitude
190° (bottom right) to 220° (bottom left), and galactic latitude −30° (bottom)
to 0° (top). (Leiden Observatory)

In optical spectroscopy the light is sent through a prism which causes
light of different wavelengths to be refracted by varying amounts. This
allows the light spectrum to be spread out and displayed so that its nature
can be studied closely. For example, each element in a star's atmosphere
produces a dark absorption line at a unique wavelength (or set of wavelengths) known from laboratory studies. An example of an optical spectrum was shown in Figure 7.1. Note that an optical absorption spectrum
is formed much as was described in regard to Figure 17.11, except that
instead of a background radio source a star is involved, and the absorption
is produced in the star's cool atmosphere or in intervening clouds in
space.

Radio waves cannot be bent in a prism, but their intensity as a function
of wavelength can be studied by tuning the radio receiver across the

TABLE 19.1 Interstellar Molecules.

			2-atom molecules			
CH	CN	CH$^+$	OH	CO	H$_2$	CS
SiO	SO	NS	SiS	C$_2$	NO	HCl
	NaCl	AlCl	KCl	AlF	PN	

			3-atom molecules			
H$_2$O	HCO$^+$	HCN	HNC	OCS	H$_2$S	C$_2$H
N$_2$H$^+$	HCO	SO$_2$	HCS$^+$	SiC$_2$	H$_2$O$^+$	C$_2$S

			4-atom molecules			
NH$_3$	H$_2$CO	HNCO	H$_2$CS	C$_2$H$_2$	C$_3$N	HNCS
HOCO$^+$	HCNH$^+$	C$_3$H	C$_3$O	HCNH$^+$	H$_3$O$^+$	C$_3$S

			5-atom molecules			
HCOOH	HC$_3$N	CH$_2$NH	NH$_2$CN	H$_2$CO	C$_4$H	SiH$_4$
			C$_3$H$_2$	CH$_2$CN		

			6-atom molecules			
	CH$_3$OH	CH$_3$CN	NH$_2$CHO	CH$_3$SH	C$_5$H	HC$_3$HO

			7-atom molecules			
CH$_3$C$_2$H	CH$_3$CHO	CH$_3$NH$_2$	CH$_2$CHCN	HC$_5$N	C$_6$H	

		8-atom molecules	
	HCOOCH$_3$	CH$_3$C$_3$N	

			9-atom molecules		
	CH$_3$CH$_2$OH	(CH$_3$)$_2$O	C$_2$H$_5$CN	HC$_7$N	CH$_3$C$_4$H

	10-atom molecules	
	(CH$_3$)$_2$CO	

	11-atom molecules	
	HC$_9$N	

	13-atom molecules	
	HC$_{11}$N	

waveband at which the molecular spectral lines are expected. Just as an FM or AM radio reveals a number of stations as the tuning is changed, so interstellar molecules act as a large number of different radio stations, each with its own call sign. In practice, the study and identification of the molecular transmissions requires expensive and sophisticated radio receivers and antennas to tune into the signals.[1]

The Production of Spectral Line Emission from the OH Molecule

To obtain a sense of how molecules create spectral lines, consider the case of OH (hydroxyl), shown in Figure 19.3. In its lowest energy condition, its ground state, it is capable of producing a set of four spectral

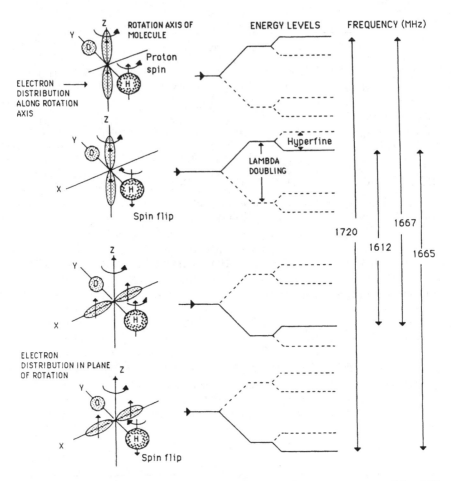

FIGURE 19.3. A schematic diagram illustrating how four energy states of the OH molecule interact to give rise to four spectral lines around 18.cm wavelength at the frequencies indicated. See text.

lines which, when observed together in the direction of a nebula or a dust cloud, unambiguously reveal the existence of OH in space.

The OH molecule consists of a hydrogen locked to an oxygen atom. The hydrogen nucleus (the proton) is glued to the oxygen by the action of its electron, which settles into an orbit about the dumbbell created by the proton and oxygen, as shown. The dumbbell molecule can rotate in one of two ways with respect to the electron location (or distribution, as it is called) and each of these states contains a different amount of energy.

Consider a simple metaphor. Imagine spinning on your toes as a ballerina does. You would be rotating around a vertical axis and this obviously requires a certain amount of energy. Now imagine a gymnast turning handsprings (which is spinning about a horizontal axis). This requires a lot more energy. These two rotation conditions are different "energy states".

Whenever two or more energy states are possible in a molecule's motion, transitions between them may occur. This means that the molecule shifts its motions from one state to another. But because different energies are involved, energy is either radiated as an emission line or produces an absorption line at a signature wavelength, as was discussed in chapter 16. Molecules may be induced to alter their energy state by collisions with other molecules or by absorption of radiation.

In the top two configuration in Figure 19.3, the electron distribution (the region where the electron spends most of its time) is along the axis of rotation, while in the lower two the electron distribution is in the plane of the rotation. These two motions contain different amounts of energy. When the molecule is caused to change its state of motion the energy difference is either radiated away, if the transition is from the higher to the lower energy state, or absorbed if the reverse is true. This transition is referred to as the lambda doublet.

In the case for the hydrogen atom (Figure 16.4), the electron had spin which is either in the same direction as or opposite to the spin of the proton. This sets up the condition for hyperfine splitting and applies to both the lambda doublet states for OH, thus giving the molecule four possible energy levels. These are indicated in Figure 19.3. Transitions are allowed between certain energy levels, and this gives rise to four spectral lines with frequencies of 1,612, 1,665, 1,667, and 1,720 MHz (around 18-cm wavelength in the radio spectrum). Under typical condition these four lines are expected to exhibit relative intensities of 1:5:9:1. This is because at any given instant nine out of every sixteen molecules in a large cloud will be undergoing the energy transition involved in the transition giving rise to the 1,667 MHz line, which will be 1.8 ($= 9/5$) times brighter than the 1,665 MHz line since fewer molecules are undergoing that transition. (The rules determining these transitions form part of quantum physics.)

The presence of OH in space is recognized from its unique spectral signature, and close study of the relative intensities of the four lines gives information about physical conditions in the OH cloud. If such a cloud also contained neutral hydrogen it might simultaneously emit the 21-cm line.

Radio Discoveries

In 1963 the radio signature of interstellar OH was discovered by the MIT group led by Alan Barrett. The four OH lines were seen in absorption against the radio source Sagittarius A. The data revealed that OH was located in many clouds between the galactic center and earth, clouds already known to contain hydrogen gas. The four spectral lines of OH were predicted to have intensities in the ratio 1:5:9:1, and the early observations confirmed this.

Then, in 1965, three research groups discovered a spectral line at 1,665 MHz which was completely unexpected. It was at least fifty times brighter than any possible OH emission at the other frequencies. A group of radio astronomers at the University of California, Berkeley, considering that they might have discovered the radio signal from a new molecule not known on earth, tentatively labelled it "mysterium," bearing in mind Huggins's "nebulium." Within months, however, the mysterium signal's intensity was found to vary in time, and then other sources were found which showed one or more of the other OH spectral lines anomalously bright. It was then realized that the strange signals were not so mysterious, but due to OH clouds in which, for some reason, the distribution of the molecules among the four permitted energy levels (Figure 19.3) was extremely peculiar. The radio astronomers had, in fact, discovered a *maser* in space.

Masers

The acronym MASER refers to Microwave Amplification by Stimulated Emission of Radiation. Microwave amplification refers to the amplification of wave at short radio wavelengths, or microwaves. Stimulated emission of radiation refers to an interesting phenomenon which can be understood by referring back to the four energy states for the OH molecule. If more than the normal fraction of the molecules were *forced* into one of the higher energy states, then transitions from that level to a lower one would be heavily favored because so many more molecules would be available to make that specific jump. This causes a line that is brighter at one frequency than predicted if one assumed conditions in the cloud to be normal.

In a vast cloud of interstellar OH, the number of molecules in any given energy state is determined by the likelihood (or probability) that the molecule contains a certain amount of energy. A predictable fraction of the molecules will always be in each of the four states, which gives rise to the relative line intensities of 1:5:9:1 mentioned above. In this "normal" case the molecules are said to be in equilibrium. The equilibrium probabilities can be calculated or measured in the laboratory. However, it is possible to force many of the molecules into a preferred energy state by *pumping* some other form of energy into them. Pumping literally "excites" disproportionate numbers of molecules to adopt a particular energy state. Every time a transition from a preferred state occurs the energy lost is quickly replenished by the pump, which returns the molecule to its starting point. (Note: A *laser* operates on the same principle.

Molecules in a crystal, such as ruby, can be pumped with energy from a power source. The crystal gets rid of this excess energy through emission of an optical spectral line, that is, light of a specific color, which produces the laser effect.)

An OH maser is produced when a disproportionate fraction of the OH molecules are pumped to higher energy levels. This may occur if the gas is bathed in an infrared glow which causes energy to be absorbed in such a way as to anomalously populate the four levels shown in Figure 19.3. Initially the energy is absorbed by the molecules and raises them out of the ground state to a higher energy level than discussed here. When the molecules then lose their energy they return to the ground states, but now populate the levels in a manner that is different from the so-called equilibrium condition.

The mysterium emission was thus identified as being due to pumping of infrared energy into a cloud of OH molecules and not to a new form of matter. These clouds are regions of star formation and within them young, luminous red stars generated high levels of infrared radiation. This energy was absorbed by the OH and re-radiated at a favored frequency to produce the peculiar spectral line.

Interstellar OH molecules can be pumped to higher energy states by either infrared radiation from stars or collisions between particles in the clouds. Interstellar OH masers were subsequently found to be common in the Galaxy. Other molecules showing the maser effect include water, silicon monoxide (SiO), formaldehyde, and methyl alcohol.

The Rat Race

Five years after the discovery of OH a research group at the University of California, Berkeley, headed by the Nobel Prize winner Charles Townes, built their own receiver and telescope to operate close to a 1-cm wavelength and discovered both interstellar ammonia and water. The water "line" turned out to be the strongest maser, as well one of the strongest radio sources in the sky. Even primitive radio telescopes and receivers such as had been used in the 1950s could have picked up this strong water line, because back then its wavelength was already known from laboratory measurements. But during the 1950s no one dreamed that water molecules existed in interstellar space and no one would have been harebrained enough to search for them!

The astronomical community was electrified by the news of this discovery, but not so Dave Buhl and Lew Snyder at the National Radio Astronomy Observatory. They had just been prevented from using the 140-foot telescope to search for interstellar water several months earlier, following technical disagreements with a referee on the proposal they had submitted. The thrill of discovery is not always easy in coming. It may even be denied an astronomer because an idea for a worthwhile experiment on a major telescope is judged unworthy by an anonymous peer. If the referee calls "out," the astronomer may not get to have his day in the sun. So it was with the detection of interstellar water vapor, a saga which, in the words of one of the participants, was "a shameful bit of scientific history that played some role in wrecking the health of one person, wrongfully rewarded other people and, to add insult to injury, is usually incorrectly remembered."

Buhl and Snyder suffered an unfortunate obstacle in the way of their exploratory program to search for interstellar water. Proposals for use of telescope time have to be framed in terms of what is already well known. Thus the justification of an exploration of the unknown is made virtually impossible. (This difficulty particularly besets those who wish to perform radio searches for signals from extraterrestrials; the search for ET is not a scientific experiment, but pure exploration.)

The astronomer can only search for what he *expects* to find. In addition, the expectation has to be described in words that a busy and anonymous peer can understand! This poses other hazards, because the ideas of a pioneer are often misunderstood, even when well posed. So if the astronomer seeking time to perform a dramatic new experiment fails to convince the referees, observing time may not be granted. Research astronomers must live with this and also learn how to phrase their requests so they are not too outrageous. The system thus encourages adherence to a code which occasionally holds back exciting new discovery. (The system of refereeing observing proposals had to be developed because time on the largest telescopes is so valuable and in such high demand.)

Buhl and Snyder had initially planned to search for radio spectral lines at a 1.2-cm wavelength from water in the Venusian atmosphere. They wondered whether there might be water in space, but could not be sure why, or where to look, except in the direction of a variety of dark and luminous nebulae. They could not have guessed that water maser emission was being produced in star forming regions and would be so easy to find.

Today radio astronomers searching for new molecules have the benefit of a short list of interstellar clouds where the molecules are most likely to be discovered. These have therefore become the best-studied clouds, the ones everyone quickly looks at in search of new molecules, and so on. But in 1968 only OH and ammonia had been found and no one knew quite where to point the telescope in order to find molecules except to look where OH and ammonia had been seen.

The unfortunate delay experienced by Buhl and Snyder therefore led to the Berkeley group's discovering the interstellar water first. To make up for this mishap, they were given priority in their next search, for interstellar formaldehyde (embalming fluid—H_2CO), a molecule whose wavelength (around 6-cm) was well known. The 140-foot telescope at the

NRAO was ideally suited for detecting this molecule, should it exist between the stars. Coincidentally, two other radio astronomers, Ben Zuckerman and Pat Palmer, were scheduled to observe with the same receiver immediately following Buhl and Snyder. They had planned to observe spectral lines from ionized hydrogen seen in emission nebulae—recombination lines. When they heard about the discovery of water and ammonia in space they thought it would be worth searching for formaldehyde, because its primary spectral line was in their observing wavelength range. They did not know of Buhl and Snyder's request, for such secrets are usually closely guarded.

However, protocol demanded that Zuckerman and Palmer inform the observatory directorate of their wish to change their experiment a little. Telescope time is given only for the defined experiment, to avoid problems arising between competing groups.

Astronomers, like any other form of homo sapiens, do not take lightly to incursions on their territory, especially in the realm of ideas. Buhl and Snyder did not like the idea that the other two might be able to observe what they hoped to detect first. However, they only had a few days of observing time, while Palmer and Zuckerman had three weeks. The first duo might discover formaldehyde, but the latter would "clean up" by observing in many directions. Thus they hoped to discover a great deal about the physics and chemistry of this molecule.

A simple compromise was reached. All four teamed up. Then, as soon as they began to observe, they quickly discovered the new molecule in space. But the observations produced a stunning surprise, as so many new discoveries do. The telescope had been pointed at a dark marking, a dust cloud, but instead of finding an emission line they saw the molecule in absorption. On the surface this is not surprising, since absorption lines in radio astronomy were common, provided the interstellar cloud is observed against a background radio source (Figure 17.11). But they weren't pointed at a background radio source! The telescope had been pointed at a dust cloud behind which no radio source was located. So what was being absorbed?

They had discovered an *inverse* maser. Instead of the levels in the formaldeyde molecule being excessively populated by pumping, they were depleted so that the formaldehyde did not radiate energy but was seen in absorption against the background radio source formed by the 3K radiation from the Big Bang! The formaldehyde inverse maser was acting as a refrigerator so that the cloud appeared colder than the universe all around it! That is why they saw absorption against empty sky!

An Avalanche of Discovery

The discovery of water, ammonia, and formaldehyde in 1968 and 1969 precipitated an intense search for new interstellar molecules. The race to detect new ones was marked by tough battles for time on the world's largest radio telescopes. Several astronomers sensed that their hunger for the thrill of discovery could be partially satisfied by finding a new molecule. Others believed that their reputations would be established forever if they could add to the stable of interstellar species. This opinion was not shared by everyone, especially those uninterested in molecules, who proceeded to witness the madness with amused detachment.

For years the bonanza of discovery continued, so that by 1973 twenty species of interstellar molecules had been found. The molecular line observations created a whole new branch of science—astrochemistry. Chemists began to take an interest in astronomy and some radio astronomers learned chemistry. In 1981 the most complex interstellar molecule found to date, the nameless $HC_{11}N$, containing 13 atoms, was detected. By the end of 1986 the number of molecular species between the stars had risen to seventy-six (Table 19.1). Most of these are organic in which the carbon atom establishes the foundation of the molecular structure. Table 19.1 reveals that the chemistry of interstellar matter is basically organic; that is, it depends on the presence and properties of the carbon atom.

Astronomers and biologists want to know if more complex molecules, such as amino acids, purines, or pyrimidines, are also present in space, especially because amino acids have been found in meteorites. Glycine, one of the common amino acids, has many energy levels that should be populated in space, but astronomical searches have not found it yet. Urea has been sought but not found. But the discovery of PN in 1986, a phosphorus-containing molecule, raised the potential that life-giving combinations may be more likely than previously expected.

Most of the molecules listed in Table 19.1 have been identified on the basis of a few spectral lines, while others have shown their presence at dozens of different wavelengths. In addition, about 150 spectral lines of unknown origin have also been discovered using radio telescopes, spectral lines no doubt due to complex molecules, although no one yet knows which ones. The large number of the unidentified spectral lines may involve at least fifty different molecules.

In the early days of molecular studies astronomers could look up expected wavelengths in tables published by the National Bureau of Standards. Today the search is slower because the required laboratory work still has to be done and is often extremely difficult. All the easy stuff was done years ago! It is possible for chemists to predict the spectral signature of a molecule, but such predictions have to be tested in the laboratory before the radio astronomer can hope to obtain telescope time and hopefully positively identify the existence of that molecule in the depths of space. In principle, with larger and better radio telescopes it may ultimately be possible to identify 150 to 200 additional molecules in interstellar space before so many spectral lines are detected that the radio spectrum becomes a toothcomb of weak lines that blend to produce a meaningless jumble of confusion.

Asymmetric and Symmetric Molecules

In general, interstellar molecules radiate spectral lines when their energy states are changed, usually by collisions. Depending on the nature of the molecule, its energy levels are determined by a combination of rotation, vibration, or hyperfine states.

Most interstellar molecules are asymmetrical in shape. For example, an oxygen atom and a hydrogen atom in OH forms a dumbbell, with the large oxygen and small hydrogen atoms glued together by an encircling electron that originally belonged to the hydrogen. Such molecules can adopt many possible energy states due to different rotational or vibrational modes, or the spin states discussed before. Many complex interstellar molecular species have dozens of possible energy states, depending on their individual architecture. The signature spectrum of such molecules may contain hundreds of unique "lines" located over a wide range of wavelength, whether optical, radio, infrared or ultraviolet.

The most important interstellar molecule is molecular hydrogen (H_2), believed to constitute over 50% of the molecular mass in the Galaxy. However, it remains essentially invisible. The problem is that it does not emit spectral lines in ways that have been discussed above because the molecule is symmetric (consisting of two identical hydrogen atoms). Cold H_2 has no differentiated energy states and therefore no observable transitions occur to produce radio or optical spectral lines. Molecular hydrogen is further unobservable because the spectral lines it does produce (in a heated state) lie in the ultraviolet and are absorbed by the earth's atmosphere. These can be seen from spacecraft and have been studied to a limited extent.

It is now known that molecular hydrogen actually predominates over all other gaseous components in interstellar space. Its presence is inferred from the widespread distribution of carbon monoxide (CO). This molecule can only radiate as much energy as it does if its energy levels are being populated by collisions in a high density medium. The only candidate molecule to provide the high density is molecular hydrogen. About 10^4 times as many H_2 molecules as CO molecules are believed to exist in the dark clouds and 100 times as much CO as other molecules. The CO is excited (caused) to emit radiation in the mm-wave window by collisions with the abundant H_2, a theoretical prediction borne out by observations. Thus CO has become a surrogate tracer for H_2, and it is now the most widely studied of all the interstellar molecules. It reveals the deepest physical secrets of the dark clouds as well as tracing galactic structure.

Note

[1] Verschuur, G.L. (1987) *The invisible universe revealed—The story of radio astronomy*, New York: Springer-Verlag.

CHAPTER 20

Formation of Molecules

The Nature of the Interstellar Medium

A gigantic interstellar jigsaw puzzle has slowly been uncovered and pieced together during the last two decades. The pieces contain information about dust grains, molecules, emission nebulae, infrared stars, supernovae, protostars, and newly formed stars. As these are assembled the overall picture is gradually recognized and the next pieces fitted just a little more readily. The view that confronts us concerns the entire process of star formation in a series of colorful images that stretch from the growth of an interstellar cloud to the appearance of a star and its planets.

The patterns on the jigsaw puzzle can be better understood by first looking at the background, the variety of forms of matter that exist in interstellar space, each with its unique properties, each interacting with the others and playing a role in determining the overall evolution of the Galaxy and its contents. We will do no more than sketch a few of the important pieces in the puzzle, because to describe the discovery of each phase, or refer to the enormous amount of work that has gone into recognizing its existence, including satellite, rocket, and telescopic observations by hundreds of astronomers, is impossible in this book.

Space between the stars contains cosmic rays, magnetic fields, atomic hydrogen (HI), molecular hydrogen (H_2), ionized hydrogen (HII)[1], molecules, and dust grains. These constituents are found in varying proportions in different regions that are very distinct from each other.

A very hot *coronal gas,* consisting mostly of ionized hydrogen, is widespread and exists between the dust and gas clouds of the Milky Way. This medium may extend to a few thousand light-years above the disk of the Galaxy. It is called coronal gas because its temperature, which ranges from 100,000 to a few million degrees K, is the same as in the corona (atmosphere) of the Sun. With a density of from 10^{-2} to 10^{-3} cm^{-3} the coronal gas makes up as much as 20 to 50% by volume of the interstellar medium. Its properties are known from observations of x-rays and also of ultraviolet radiation from ionized oxygen atoms (OVI) made by the Copernicus satellite.[2] Very hot regions with bubble-like shapes up to 30 pc across, and temperatures as high as 10^7 K have been found around very hot, young stars, which gives the interstellar medium the appearance of Swiss cheese.[3]

Diffuse interstellar clouds, containing mostly atomic hydrogen (chapter 17), of density as low as 0.02 cm^{-3}, but usually from 5 to 100 cm^{-3}, and with temperatures from 20 to 200K, are found within an *intercloud gas* consisting mainly of neutral and ionized hydrogen with densities ranging from 0.1 to 1 cm^{-3} at a temperature around 2,000 K. The intercloud gas is very difficult to observe, although interstellar absorption lines produced by ionized atoms of sodium and calcium have been observed from this medium. It was in diffuse clouds that molecules such as OH, formaldehyde (H_2CO), and carbon monoxide were first discovered.

These clouds are cooled by radiation lost by small amounts of ionized carbon atoms that exist within them, and if the temperature is greater than 100K the heat is lost by radiation from molecular hydrogen. At the low densities these clouds are not expected to contain molecules.

Very close to the central plane of the Galaxy, in the Milky Way itself, are found two distinct types of molecular clouds. *Small dark clouds* or *small molecular clouds*, a few light-years in diameter, with densities 10^3 to 10^4 cm^{-3}. These are the sorts of objects photographed by Barnard, for example, in the vicinity of the star ρ Ophiuchi (Figure 4.3), which contains a few hundred solar masses of cool material at a temperature of 10 to 20K. Cores as cold as 5K, with densities 10 to 100 times greater than in surrounding envelopes, exist within the small dark clouds. These objects contain mostly molecular hydrogen. Since there is no internal starlight to heat these clouds they become very cold through radiation lost from CO molecules. Any heat they possess may come from gravitational collapse or cosmic rays striking atoms and molecules within them.

Globules are isolated dark clouds a fraction of parsec across, with very high density such as can be seen in Figure 8.6. They are similar to the cores of small dust clouds but with the envelopes of less dense matter missing.

Giant molecular clouds (GMCs) are the most massive entities in the Galaxy, containing up to a million solar masses of mostly molecular hydrogen, as well as enormous quantities of carbon monoxide and dust and riddled through with many species of organic molecule (listed in Table 19.1). Typical sizes are 20 to 200 pc across. Overall densities are as high as 10^7 cm^{-3}. When their average densities reach above 10^3 cm^{-3}, conditions for star formation are created. A GMC is usually surrounded by an enveloping cloud of atomic hydrogen, which is absent inside the cloud. It will contain one or more dense cores a few dozen parsecs across, each of which is a possible site for star formation. The GMCs are at about 10K over most of their volume although newly born stars can warm them to 70K over a few percent of their volume.

These molecular clouds contain half of the mass of the interstellar medium, and fill only 2% of the volume. Yet most of the interstellar dust, as well as many of the most complex molecules, are found in these objects. One of these clouds lies immediately adjacent to M17 (Figure 19.1). Another lies behind the Orion nebula (Figure 1.4) and forms part of a huge swath of molecular clouds outlined by their CO emission, as is shown in Figure 20.1.

The most dramatic one known is invisible. It lies behind vast quantities of dust close to the galactic center, specifically in front of a radio source known as Sgr B2, an HII region. This GMC contains vast amounts of matter, between 3 and 5 million solar masses of mostly molecular hydrogen, which is riddled with the majority of the molecular species listed in Table 19.1.

GMCs are nearly always associated with clusters consisting of O and B stars, the youngest, hottest and most massive stars known.

Hot molecular cores in GMCs have been detected by radio interferometers capable of seeing great detail. These cores have temperatures of 2,000K, densities as high as 10^8 cm^{-3}, and contain 50 to 1,000 solar masses of gas.

The GMCs as a class are found along the Milky Way, and often the outlines of the CO clouds defining relatively nearby GMCs agree perfectly with visible dust clouds. The GMCs were first discovered by Pat Thaddeus, Marc Kutner, and Ken Tucker from Columbia University using the 5-m radio telescope of the University of Texas, Austin. Subsequently the Columbia University radio astronomers built a special purpose CO telescope on the roof of a building on the Columbia campus; this despite the electrical interference produced by the city (itself permeated by carbon monoxide) which at the CO wavelengths of 8 mm was not yet so severe as to make the task impossible.

The majority of the GMCs, of which 4,000 may exist in the Galaxy, are found between 4 and 8 kpc from the galactic center. Star formation activity therefore seems to be concentrated in the region of the Milky Way within the sun's distance from the center (8.5 kpc).

The Formation of Interstellar Molecules

Interstellar molecules are tracers that tell about the physical conditions in dark clouds and the chemistry occurring in their environment. Astronomical detectives who attempt to solve the mystery of molecular formation have several specific theoretical clues to aid in the interpretation of their observations.

1. They know that molecules are quickly destroyed when exposed to ultraviolet radiation in open space. Molecules will generally not last 100 years in space unless shielded from the UV. Therefore, molecules must form in dark clouds, or dense circumstellar envelopes surround-

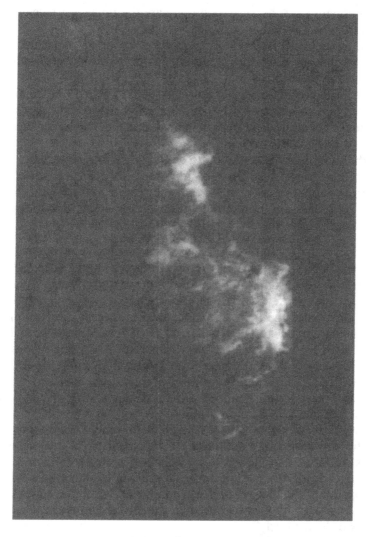

FIGURE 20.1. An image representing the total amount of CO molecular line emission in the region of known as Orion B. The extent of the image is 4° vertically and nearly 3° across (centered on approximate right ascension $5^h 48^m$, declination $-1° 30'$). The region is located north of the Orion nebula (Figure 1.4) which also lies in CO clouds such as are seen here. The Horsehead nebula (Figure 5.2) is seen as a small, hooked structure on the lower right hand edge of the main band of CO emission. The bright spots that lie just above the Horsehead are two nebulae known as NGC 2023 and NGC 2024. A remarkable infrared image of the latter is shown in Figure 21.7. The dark rectangular patches along the left and right side of the image have not yet been fully mapped. Nevertheless, the CO is clearly distributed in filamentary structures reminiscent of interstellar cirrus, e.g., Figure 11.4. (John Bally)

ing certain stars, because only there are the molecules shielded from destructive UV radiation.

2. The range of the isotopes observed for a variety of molecular species shows that the molecules could not be formed in the atmospheres of old, evolved stars.

3. The presence of molecular ions in space (Table 19.1) implies that formation sometimes occurs by interaction between the gaseous versions of the molecules rather than between solid, frozen particles which could not become ionized.

4. Laboratory work, attempting to simulate conditions in space, shows that dust grains must play an important role for some molecules in forming a matrix upon which they may gather and combine. Some of the molecules are formed only in envelopes of gas ejected from old, cool stars, or in the prenatal stages of a star's life.

It appears that ion-molecule reactions which involve collisions between only two bodies are some of the most important and occur when the temperatures are from 10 to 100K, and the densities range from 10^2 to 10^7 cm^{-3}. The process works as follows: Cosmic rays penetrate the dark cloud and ionize the most abundant species present. For example, H_2 becomes H_2^+ and H^+. These two, and He^+, react quickly with H_2 and CO to form H_3^+ and C^+. H_3^+ is very important in driving the reactions, a fact that has only recently been recognized. This molecule then combines with CO, O, N, and H_2 to form new products such as HCO^+ and NH_3^+, and they, in turn, combine with other atoms, such as metals, which may be present.[4]

Barry Turner, at the National Radio Astronomy Observatory, an internationally recognized authority on the chemistry of interstellar molecules, sums up as follows: "A complicated network of these kinds of reactions constitutes the ion-molecule scheme of interstellar synthesis. Application of the theory involves complicated computer codes that entail as many as 1,500 reactions that are solved numerically . . . ."[5] (In other words, tediously!)

The details are far from understood and complications are introduced because of the way time-dependent variations at each step affect the outcome. Yet the process of ion-molecule reactions is well-enough understood to account for the smaller (four atoms or less) molecules in space.

A second process, of taking atoms and joining them to form molecules, or combining simple molecules to form more complex ones, often involves dust particle surfaces. The nature of the grains is only fairly well understood (chapter 10), but they are known to be solid objects, possibly based on a silicon core surrounded by water ice infested with a huge variety of organics as well as complex, as yet unidentified, molecules.

Four steps appear to be important in all formation processes involving grains:

1. The atom or ion must stick to the surface of the grain. This is called *adsorption*. For example, almost all molecular hydrogen that collides with a grain will stick to the surface, while 20% of the visiting hydrogen atoms also stick. Heavier atoms are more choosey and stick less frequently.
2. The atom or ion has to be able to migrate across the surface to meet its mates.
3. Coupling between fragments must occur. For example, the observed abundance of interstellar H_2 cannot be explained without invoking the presence of a comfortable solid surface on which the hydrogen atoms can "mate in holy matrimony" to form H_2.
4. Finally, the product of any such interaction—the new molecule—has to escape or be ejected into space or we would not be able to observe it from the earth.

Details of these molecular formation steps depend on temperature, composition, and physical properties of the grain surfaces.

Grains are covered in ices in cold clouds, but not in warm clouds. A large variety of organic molecules can be synthesized on silicate surfaces, including the HC_nN series (Table 19.1) while hydrocarbons are formed easily on graphite (carbon) grains. Polymers should also form, and are hard to dislodge unless heated. Metallic oxide grains can form H_2O, NH_3, H_2CO, and HCO very efficiently. When ejected by heat they will rapidly photodissociate to form CO, OH, and NH, which are among the most abundant of all the observed interstellar molecules.

The grains are also sites where otherwise hard-to-form species can shelter from the winds of space. They can survive in hiding until the grain enters stormy interstellar space and is torn apart.

Although grain surface chemistry has many difficulties, it may account for some molecules which cannot be explained by gas-phase, ion—molecule reactions. In cold clouds the grains may not play a role at all, while in warm clouds they may augment the other processes.

The total disruption of grains may be a source of molecules in many dust clouds rather than having them produced by gradual release from grain surfaces.

An important concept in understanding interstellar molecule formation is known as equilibrium chemistry. If a large cloud of atoms and molecules is held closely together (as in a flask or a container of some sort) the molecules will undergo a series of chemical interactions. After a while the gas will contain a certain fraction of a wide variety of species, determined by the continual balance between formation and destruction of molecules within the mixture. At this point the mixture is said to have reached equilibrium. In the laboratory, equilibrium chemistry occurs in any liquid or gas in which a variety of molecules are mixed and allowed to react with one another. In space this process may occur in the envelopes surrounding cool stars, where a tremendous density of gas is gathered and temperatures are sufficiently low to allow equilibrium chemistry to proceed. In the actual atmospheres of the cool stars the temperatures would be too high to allow molecules to be formed, or survive for any length of time.

Chemical (or thermochemical) equilibrium is important for forming molecules in the high-density, high-temperature regions in circumstellar shells around certain stars, such as cool red giants at the end of their lives, which eject large quantities of matter. These were discussed in chapter 10 with regard to grain formation.

A specific example of such a star is IRC 10216, a star containing a huge fraction of carbon, which is observed as a strong infrared source hidden behind a dense dust cloud. In some envelopes all carbon is in the form of CO. It appears that such stars are surrounded by a series of shells of material with very different properties. The inner shell contains many molecules, in particular some of the heavy diatomic species such as the recently discovered NaCl and AlCl. Densities may be as high as 10^{10} cm^{-3}. This layer is surrounded by another where solid grains are forming. The next layer is one where other molecules have condensed onto the grains. The outer layer is particularly rich in certain molecules, such as cyanoacetylene. Beyond that the circumstellar shell then becomes exposed to starlight from deep space and the molecules and grains are destroyed. These stars are not believed to contribute to the molecular population of the general interstellar medium.

Shock waves striking clouds may also facilitate the formation of molecules. Shocks are supersonic compressions that occur when the velocity of the gas exceeds the local sound velocity and are produced by the expanding remains of exploded remnants. Figure 20.2 shows part of an interstellar dust ring which indicates the existence of an old supernova remnant where shocks are likely to be present.

Jets emerging from protostars (see chapter 21), expanding HII regions, stellar winds from O and B stars, supernova explosions, cloud–cloud collisions, and the spiral density wave that produces spiral structure all generate shocks. These compress the gas and heat it to 2,000 K and their presence is observed in broad molecular lines seen in many GMCs.

A basic principle related to this phenomenon is that heating of the gas by the passage of shock waves overcomes the energy barriers that prevent chemical reactions. This is the energy that holds molecules together and has to be overcome in order to take the molecules apart again so as to rearrange their structures. Shock chemistry, for example, explains the high abundance of interstellar CH$^+$, which is 100 times more abundant than ion–molecule chemistry permits. It also explains the existence of SiO found in high-velocity flows from protostars (see chapter 21).

Molecule formation and destruction processes involve a variety of interactions between atoms, ions, ionized molecules, free radicals, solid

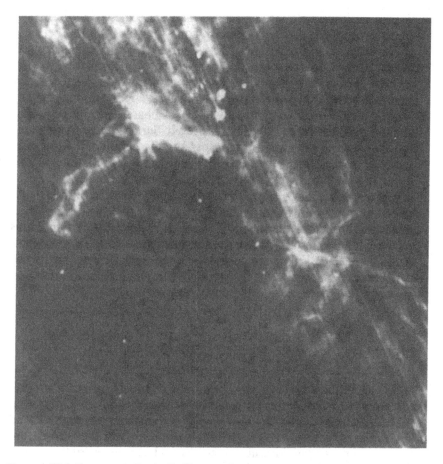

FIGURE 20.2. Segments of a shell of interstellar dust revealed by the IRAS satellite. This cirrus shell, visible at a 100 micron wavelength, is probably related to the remnant of a very old supernova and indicates a region of space in which shock waves are present. Such shock waves are conducive to the formation of complex interstellar molecules in these interstellar dust clouds. The center of the filamentary arc is located at approximately galactic longitude 140°, latitude 35°, the size of the entire field being about 15° across. The stripes are an artifact in the original data. (IRAS Plate 6, Infrared Processing and Analysis Center, Caltech)

dust grains, cosmic rays, stellar radiation, and shock waves. In order to understand the phenomena related to molecule formation, not only does the astronomer have to contend with uncertainties in the clumping and physical conditions in the cloud, but if shocks are present they will contribute more than a little to molecule formation processes.

In summary, there are three basic places to form interstellar molecules.[6]

1. Within molecular clouds. As diffuse material coalesces into a cloud, densities increase so that two-body collisions can occur. Then gas-phase reactions and/or reactions on the surface of grains can take place, which, in turn, react to form more complicated molecules.
2. In high-density, high-temperature regions near young protostars, or stars and presolar nebulae embedded in dense clouds. Here the density may be high enough so that many-body gas phase reactions can occur and thermochemical equilibrium may result. Surface reactions on grains can also occur. In the case of mass loss and disruption of the placental material by the stars, molecules are ejected into space and appear to occur in star-forming clouds (or non–star-forming clouds).
3. In gaseous envelopes of cool, evolved stars independent of interstellar clouds. Temperatures and densities are high in these shells and molecules are formed in thermoequilibrium. It is here that certain grains are formed (chapter 10). When the star loses mass the grains and molecules are ejected into space, where the grains may be destroyed when they become exposed to the background ultraviolet radiation that permeates all of space outside the dusty regions.

How Do We Know How Many Molecules Are Involved?

The essence of the answer to this question is relatively simple. Astrochemists use what is known as an equation of transfer—which is just what the label implies. It describes how energy is transferred through a cloud of molecules (or atoms). To relate this equation to "real life" the astronomer has to know something about the physical parameters of the source. The observations involved in interpreting the equation relate to the nature of the spectral line observed: its brightness and its width. The probability that a given molecule will undergo a transition in a given time can be calculated, or is known from laboratory measurements, and this allows the equation describing how much energy is radiated to be worked out. This, in turn, depends on the frequency of collisions that may be occurring, or the amount of energy impinging the cloud in the form of cosmic rays or UV, for example. The probability calculations are applied to an ensemble of molecules at a certain temperature, and in a specific radiation field which allows the expected spectral line brightness and width to be predicted. Comparison with observations then allows the reverse process, the derivation of how many molecules are involved in producing the observed signals, to be calculated.

The transfer calculations are made far more difficult when it is recognized that the molecular distribution in the cloud may be clumped. If one clump is very dense it will not only emit radiation, but will absorb energy from around it. For this reason astronomers are very interested in mapping the fine-scale structure in clouds and use the world's largest radio telescopes to observe this clumping (Figure 20.1). The more detailed the knowledge they have about the cloud, the more accurately the transfer equation can be treated. To add to these difficulties, within one cloud such as a GMC—the clumping of different molecules will vary. More of one species may be present in one region while another species clumps somewhere else.

To perform the correct calculations, the temperature distribution of the cloud also needs to be known. While observations of the CO line give reliable values, these are usually for the outside surface of the cloud in question and not for the inside. If there exist hidden heat sources in the cloud the transfer equation is altered, which means that the interpretation of the observations also has to change. This produces another unknown in the equations.

If the fingerprint of a given molecular type contains a recognizable pattern involving several lines, it is possible to determine very accurately the nature of the conditions in the molecular cloud because the relative brightness of the lines emitted by various transitions is determined by what are known as the equilibrium conditions in the cloud. At different temperatures equilibrium conditions are different, and this is reflected in the spectral signature observed. For a detailed discussion of this topic the reader is referred to other sources.[7]

It is clear that as we look closer at the subject of molecule formation in interstellar space it really is very specific and complicated. Many astronomers are involved at chipping away at the specifics and will be doing so for decades to come. What they are certain of, though, is that it is the dark clouds that contain the dust and molecules that are the true stellar nurseries, a far cry from what Barnard and his peers might have expected of the vacancies between the stars.

Notes

[1] The notation HII refers to ionized hydrogen, the state when electrons have been removed from orbit about the protons, so that these two drift freely in the interstellar cloud.

[2] The notation, OVI, refers to highly ionized oxygen.

[3] McCray, R., Snow, T.P. (1979) "The violent interstellar medium." *Annual Rev. Astron. Ap.*, 17:213.

[4] The reader interested in details is referred to Turner, B., Zuirys, L. (1988) "Interstellar molecules and astrochemistry," in *Galactic and Extragalactic Radio Astronomy: 2nd Edition*, New York: Springer-Verlag, p. 200.

[5] Ibid.

[6] Ibid.

[7] Ibid.

CHAPTER 21

Molecular Clouds and Star Formation

Star Forming Regions

Manifestly the most satisfying and rewarding consequence of the discovery of molecules in space has been the enormous growth in knowledge about star formation. Just as prenatal processes in living things are hidden from view, so star birth occurs in the depths of clouds which are filled with absorbing dust grains. This means that we on earth cannot see the stars either before or during their private moment of birth. It is usually thousands of years before the placental material of the surrounding cocoons of dust are sufficiently swept away to reveal the newly born.

Yet, astronomers can "see" into the stellar wombs just as a physician using ultrasound can "see" the fetus in the human womb. But the astronomer observes radio waves and infrared radiation which penetrate the dust layers and clearly reveal just what is happening in the depths of clouds where protostars are being hatched. This fortunate state of affairs results because an abundance of complex molecules broadcast their presence from within the dark clouds. The hidden secrets of star formation are hidden no longer.

Molecular transmissions provide a beautifully sensitive diagnostic tool for the astronomer. Each molecule generates its own set of spectral lines which act like a string of Christmas tree lights. Depending on the temperature and density in a cloud, a different combination of lights is lit. Observation of a particular combination of molecular lines from a cloud gives very explicit information about its physical properties. When combined with mapping observations, which delineate the extent of the clouds, the astronomer then has virtually all the data required to make a detailed model of the cloud and the state of star formation occurring within it.

The study of star formation in the dark clouds photographed by Barnard has now become one of the hottest topics in all of astronomy. It represents a continuing saga of detective work which has uncovered examples of interstellar clouds and stars in virtually every phase of evolution somewhere in the galaxy, from the initial diffuse clouds, through molecular clouds, to globules. Even the very moment of star birth itself has been revealed.

The study of spectral lines from interstellar molecules tells so much about the physics and chemistry of interstellar processes. The physics is related to the mechanics of star formation, while the chemistry tells how the molecules are formed. Let us visit those remarkable regions of space where stars are now being born.

The Nature of Dense Molecular Clouds

The giant molecular clouds (Chapter 20) are clearly the birthplace for stars. GMCs and young O and B¹ stars are found together and define the spiral arms of the Galaxy (chapter 17). Since the interstellar dust obscures starlight, the first indication that star formation is taking place in the dark molecular clouds comes from other quarters, especially as the presence of bright infrared sources and masers within the cloud boundaries. The HII regions, such as seen in Figure 19.1, often show hot, ionized matter streaming into space. These HII regions are produced by stars recently formed near the edge of a GMC, like a blister on its surface. The hot stars eat their way into the surrounding molecular hydrogen, destroy it, ionize the atoms, and usher them on their way into space at speeds of up to several tens of km/s. The blister breaks and spills newly processed stellar stuff into space.

Infrared and millimeter radio astronomical observations of molecules, especially of CO, have revealed even more information about the GMCs, which show evidence for the formation of low-mass stars as well as the high mass O and B stars. Smaller molecular clouds show no evidence of the formation of massive stars, although clearly low-mass stars have been created. This subtle distinction has turned out to be very important to the overall understanding of star formation.

Heating and Cooling of Clouds

What really determines whether stars might form in a given cloud? The answer is a question of balance (chapter 14)

Interstellar clouds experience a variety of heating and cooling effects which determine just how much kinetic (heat) energy they will contain. This, in turn, determines whether or not they will expand and dissipate, or collapse to form stars.

A primary heating mechanism in interstellar clouds is near-collisions between atoms and cosmic rays. These interactions cause the atoms to be accelerated, so they move more rapidly. Bear in mind that the temperature of a gas is a measure of how fast the particles are moving. Cooling occurs whenever the cloud radiates away energy in the form of infrared, light, or radio waves as, for example, by the thermal emission process. Cooling also results when energy is lost in the form of spectral lines from molecules such as CO. Although each molecule loses only a tiny amount of energy in the process, the net result for a cloud containing a vast number of molecules is substantial heat loss. Molecular spectral radiation is therefore a very important mechanism in determining the balance of energies in interstellar clouds. By making the appropriate observations of molecular lines the astronomer can determine which process—heating or cooling—is dominant and whether the cloud is presently contracting (due to cooling) or expanding (associated with heating).

Collisions between interstellar clouds also creates heat. Furthermore, when hot stars are born within a cloud, their UV radiation will heat the central regions. This energy may be great enough to eventually disrupt the surrounding cloud and destroy it. The overall challenge to understanding star formation is to figure out whether the state of balance is such as to favor the collapse of clumps of gas under the influence of gravity to form stars.

A New View of Star Formation

Observations of molecular clouds suggest that they contain so much mass that as a rule gravity will overwhelm the resistance provided by all possible internal energy sources (chapter 14). The GMCs should be collapsing so fast that they must form stars very readily, at the rate of 1,000 solar masses of stars per year in the Galaxy. As is so often true when theory meets observations head-on, the predictions do not match the facts: the observed rate of star birth is about three solar masses per year. This is based on knowing the total mass of stars (6×10^{10} solar masses) and the age of the Galaxy (approximately 2×10^{10} years). Some theoreticians suggest that the birth rate is as much as six solar masses per year, of which four stay in long-lived stars and the other two rapidly end their lives as supernovae or other less violent forms of star death. It is widely agreed that there is no evidence for star formation much above these rates, which means that something must be acting to prevent the GMCs from forming stars, even if they seem primed to do so.

What is at issue is that the observed star birth rate is far less than the rate that could be supported by the GMCs, but only if they were free to collapse, unimpeded. Herein lies one of the great mysteries which has challenged terrestrial detectives in the search for knowledge on the trail of interstellar matters. What prevents the GMCs from forming more stars? Something, some force, must act to stop the GMCs form collapsing too fast, a force that acts as the great inhibitor of star birth. If this inhibitor did not act, all of interstellar matter might long since have disappeared into stars.

The question boils down to finding what other force alters the equations of balance. Rotation in molecular clouds is found to be insufficient to hold up the clouds against gravity, magnetic fields are too weak to control the movement of material in a significant way, and the temperature in the clouds is not high enough. Which leaves only one other possible culprit; turbulence. This phenomenon produces an outward directed force which, if left to itself, would disrupt the cloud. The important point is that turbulence acts to oppose gravity so the answer may be found here.

Turbulence is defined as irregular motion, chaotic, if you like. In terrestrial clouds turbulence can be seen as bits and pieces of the clouds swirl this way and that. A crowd flowing along a busy sidewalk in a major city at rush hour may show both order and turbulence as people dodge and make way for others, all striving to avoid head-on collisions. Gusty winds in a squall are turbulent. The very word conjures up feelings of discomfort.

The presence of turbulence in interstellar clouds is exhibited by the way it broadens the spectral lines from molecules or atoms. The greater the turbulent motion, the greater the Doppler shifts produced by the motions. Thus, the broader the spectral line. But since temperature also broadens the lines (chapter 16), it is necessary to distinguish between the two mechanisms. The emission of molecular spectral lines is sensitive to temperature in the cloud. That, after all, is why it is so fortuitous that molecules show the "Christmas tree light" effect. The study of the relative brightness of those lines allows the temperature of the gas to be derived. If the observed line width (Figure 16.5) is then broader than predicted using this temperature the rest of the broadening must be due to turbulent motion (provided rotation, and streaming along field lines, which are both highly ordered and produce a different signature).

Turbulence sometimes involves motion faster than the speed of sound (supersonic) in the medium or may be gentler (subsonic). The nature of the interstellar turbulence, it turns out, defines the routes that can be followed in the process of star formation.

In recent years, star observations of OH and water masers revealed that motions within GMCs are turbulent and contain sufficient energy to prevent collapse of the clouds. However, such motion should cease after a time. This occurs when the bits and pieces in the cloud have collided so often that they dissipate their energy. Because molecular clouds are known to survive for millions of years, based on observations of the way they are spread about the Galaxy and the ages of stars within them, there has to be a constant supply of turbulent energy to keep them going. If turbulence is to be invoked as the force that prevents rapid collapse it must not die away too soon. Another wonderful astrophysical mystery that appears to be elegantly solved concerns the way turbulent energy is constantly resupplied.

Galactic Rotation and Turbulence

The energy source that replenishes turbulence and prevents more rapid star formation may, according to one scenario, be the rotation of the Galaxy itself!

The Galaxy does not rotate like a solid wheel (chapter 17). If it did, then the stars and gas farther from the center would rotate about the galactic center faster than material close in simply because that material has a greater distance to travel in its path about the galactic center. Observations of most spiral galaxies show that beyond a few thousand light-years from the center matter moves at about the same speed no matter how far from the center it is.

Stars and interstellar matter orbit the galactic center at 220 km/s which means that during the time it takes for the Sun to orbit the galaxy once (240 million years) material further out is left behind, and material inside the solar circle moves well ahead of us. This is known as differential galactic rotation (chapter 17). Consequently, different regions of the Galaxy move at slightly different rates, and it is then possible for some of the rotational energy of the Galaxy to be converted into turbulent motion as clouds swirl about each other. This is especially important within the giant molecular clouds.

The phenomenon of turbulence resulting from the interaction of two streams flowing at different rates is within the experience of anyone who has ever driven on a freeway at rush hour. Where a crowded access road carries traffic onto the freeway, the smooth flow on the freeway is disrupted as drivers weave between lanes and have to slow down or speed up as they jostle for position. This is turbulence in action! Highway engineers design the roads so as to minimize the turbulence generated by interacting flows, which otherwise causes the energy of the smooth flow of cars to be dissipated in huge traffic jams!

Once large units of interstellar mass, such as GMCs, start to indulge in turbulent motion, this energy can be fed into smaller scales of size as the material moves chaotically and breaks into smaller pieces, each travelling rapidly in random directions. The same phenomenon "cascades" downward into smaller "scale-sizes," as they are called. Energy is literally poured from one scale-size to the next, from the largest (the galactic scale) to the ever smaller—GMCs, SMCs, and then dense cores.

On the galactic scale large masses of gas are torn asunder by differential galactic rotation and these collide with each other and create large-scale turbulence. Turbulent elements (blobs within the cloud) move randomly, interact, and generate smaller scale turbulence, and so on. Energy is fed into the smaller blobs, some of which may ultimately form stars. However, star formation can only begin when this turbulent "cascade" process ceases; otherwise the cascade would go on indefinitely and stellar-sized masses would never be formed.

A new condition of balance has to be attained for star formation to occur. This condition requires that small pockets of gas containing sufficient mass to overcome the disruptive effect of internal heat, but not enough turbulence to fight off gravity's pull, must separate themselves from the cascade of energy. These, and only these, masses will be able to attain stardom.

An alternate scenario, proposed by Charles Lada and Michael Margolis of the University of Arizona Steward Observatory is one in which the energy produced by a by-product of star formation, bi-polar flows (see below), injects sufficient mass and energy are injected back into the clouds so as to replenish turbulence. Thus the process of star formation gives rise, through the consequences of gravitational contraction, to effects which further assist the process. As Charles Lada put is, "Gravity is really a wonderful force."

The Point of No Return

An important point is thus reached. Segments of a giant molecular cloud can only be prepared for stardom when no more energy is fed into the mass and local cooling in the small globule is rapid enough that gravity wins over internal pressure, which tends to heat it up. Then the cloud segment can collapse; that is, gravity pulls the gas inward. Such segments, or globules as they are called, must literally separate themselves from the surrounding gas cloud. They become "decoupled" from their environment and go into isolation to evolve on their own, oblivious of whatever happens around them. No more energy may enter or they will be unable to form stars.

The decoupled globule will collapse to form a star even as the enormous molecular cloud all around it continues to be held up against gravity by the turbulent energy derived from differential galactic rotation.

This view of cascading turbulence has led to an elegant picture of star formation, but it still suffers problems. It turns out that the small clumps that decouple themselves will form only low-mass stars (with masses ranging from 1% of the solar mass to a few solar masses). How, then, do massive stars form, those which are 20 or 30 times the solar mass? Since such stars are observed, there has to be some way to account for their existence.

Observations of the giant molecular clouds suggest that coalescence of several smaller clouds, as well as the presence of shocks within the clouds, will cause enough matter to gather into larger clumps so as to encourage these masses of gas to collapse to form larger stars. Three aspects of this process must to be considered. The accretion (or coalescence) within a large cloud containing small clumps will cause more matter to gather near the clumps as the gravitational influence of the accreting matter stretches farther and farther into space. This process feeds back on itself, so that the mass grows ever larger. In addition, more mass is driven into the gravitational tentacles of the accretion mass by external triggers, such as supernova remnants or the emergence of HII regions. Both of these can move large volumes of gas, and if they happen to drive matter close to the growing pockets of gas they will be feed them an added diet of gas and dust.

When the growing mass contains enough material it will collapse to form a massive star. Not surprisingly, massive stars tend to form in the neighborhood of large masses, which help each other gather even more mass. It is in these massive cores of the GMCs that material becomes more centrally condensed with time and where the massive stars form.

The star formation scenario thus reveals that low-mass star formation peaks and then drops off when massive star formation begins. This occurs when matter is no longer syphoned off to form smaller stars, but accumulates in larger amounts. Small molecular clouds have less gas to start with massive stars cannot form within them. The massive stars, which are also the hottest ones which form HII regions such as the Orion Nebula are found only in GMCs.

Clearly we have come a long way since Barnard wondered whether the Orion nebula indicated the beginning or the end of worlds. Unknown to him, those dark objects held the story of the creation of stars.

In summary, galactic rotation is a vast source of turbulence which supports the giant molecular clouds against gravitational collapse. Only fragments within such a cloud will collapse to form stars. A GMC may initially nurse large numbers of small stars, which in turn will attract more matter and encourage the birth of massive stars. These are usually hotter and they, in turn will heat up surrounding gas, which then destroys the clouds from which the stars were formed. The newly born stars literally shatter their placental material, and the GMC breaks up into a lot of small molecular clouds which either go their own way or form more small stars.

Star Formation-A Slightly Technical Overview

The details of star formation are now so well understood that is worth offering a slightly more technical summary. This is done so as to dramatize just how far astronomers have come along the road of knowledge since they wondered whether the dark markings were holes in the heavens.

Star formation may be seen to follow two distinct paths. First, low-mass stars form from low-mass cores in molecular clouds. These are the dark clouds seen in so many of the illustrations of this book. Not all the dark clouds are presently primed for star formation, but they are all involved in this endless dance. Supersonic turbulence then creates shock compressions and further condense structures of gas into sheets, filaments, and clumps. Smaller substructures may also form within them. Eventually these structures become small enough so that the internal motions become subsonic, suffer no more shocks, and will cease to develop further substructures. This is the beginning of star formation.

Supersonic turbulence should produce stars in low-mass, self-gravitating dense clumps, provided that subsonic motions exist within the clumps. In the process a density gradient builds up. This means that there is a range of densities in the structure from the inside to the outside edge. Once this occurs the flow of turbulent energy into the clump is hampered by a density "wall," so that the flow of energy from the outside is decoupled (stopped). The remaining turbulent energy already present dissipates and is lost as heat. Energy stops entering the dense core. Now stars begin to form by gravitational collapse. This process happens in both GMCs and SMCs and low-mass stars are expected to form widely within both types of clouds, as is observed. Very cold cores form smaller stars. Warmer cores form more massive stars.

According to Barry Turner and others who have studied the star formation regions by observing their molecular lines, the classical picture of gravitational collapse of an interstellar gas cloud and fragmentation to form a cluster of stars is no longer correct, because it provides no mechanism for forming stars of different mass at different times, something that is clearly observed in the GMCs.

The observational rules for star formation may be summed up as follows: If no external shocks exist and no enhanced accretion due to formation of massive clumps occurs, then turbulence is eliminated in subsonic clumps, the small ones, where low-mass stars form. If there is enhanced accretion of matter, due to the formation of large blobs which rapidly attract more matter, then a massive core in which motion is supersonic will form substructures which will then subdivide to form low-mass stars.

When an external shock impinges on a massive clump it compresses the clump of gas.[2] Density gradients resulting from compression allow the turbulence to decouple. Gravitational collapse may ensue and lead to the formation of a massive star. The larger the initial mass, the weaker the shock needed to cause the collapse. Since the spiral density wave acts as a perpetual shock wave that orbits the galaxy, young star-forming regions are found along the edge of the spiral arms, indicating where the spiral density wave is currently situated. This is particularly obvious in face-on galaxies (Figure 17.1), where ridges of bright light from newly formed stars lie parallel to ridges of dark material, the dust within which protostars lurk, protostars that have not yet burst out of their shells.

To summarize, a global source of turbulent energy, which must originate in differential galactic rotation, couples in to support the balance of both SMCs and GMCs. The GMCs form when SMCs are driven together by the spiral density wave that constantly circulates around the galaxy, the perpetual shock wave that triggers the formation of stars at the edges of spiral arms. Both the SMCs and GMCs may then form subsonic structures which then produce low-mass stars. In addition, GMCs form accretion cores within which lots of newly formed small stars attract more gas, or coalesce from collisions between clumps. The massive cores so created contain heated gas and form only massive stars when they are shocked. These stars interact violently with the parent GMC and disrupt it. Some of the gas fragments form new SMCs, which are recycled through the next spiral density wave to make more GMCs.

The Not-Quite-Ready-for-Startime Stars

When the protostars that form in the dense cores begin to generate enough heat energy to disrupt their placental, dust-filled molecular clouds they may become visible to optical telescopes. The most recently formed objects in this class are known T-Tauri stars, named after a prototype variable star in the constellation Taurus that lies within the bounds of the dark cloud originally photographed by Barnard. These are solar mass objects whose surfaces undergo repeated explosions as matter continues to fall in from surrounding space. On its inward journey this material first spirals and seeks haven in a disk of matter around the star. There it lingers, in the halfway house known as an accretion disk, from which it may then move on to plummet onto the star's surface.

The T-Tauri stars are typically found in the cores of dust clouds and have recently been recognized in large numbers by observations made with the Infrared Astronomical Satellite—IRAS. Figure 21.1 shows the location of infrared sources within the ρ Ophiuchi dark cloud core, half of which are believed to be T-Tauri stars. According to Richard Larson[3] at Yale Observatory, a dense core in the dusty molecular cloud will collapse and form a small accreting nucleus, or embryo (or T-Tauri) star. During the infall phase radiation from the central object is absorbed by surrounding dust and heats it up, and the dust then radiates in the infrared. The other half of the infrared sources in Figure 21.1 are protostars, which are objects still contracting to form stars. They are far cooler than stars (only a few hundred degrees, with heat generated by gravitational contraction rather than thermonuclear processes, as is the case for stars.

Many such objects have been found in the IRAS survey. The entire ρ Ophiuchi dust cloud, for example, contains as many as eighty young stellar objects, almost certainly T-Tauri stars, as well as Herbig—Haro (HH) objects. The latter are wispy nebulosities, first discovered in the early 1950s, consisting of hot ionized gas, that emit lots of ultraviolet radiation. They vary rapidly in brightness because they are sweeping up

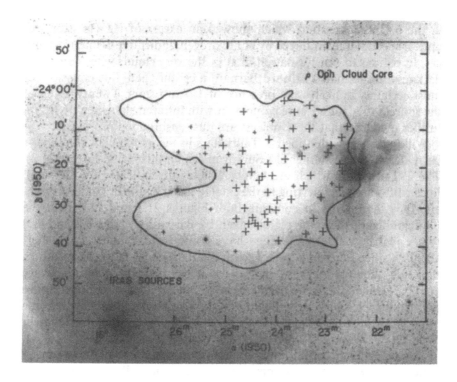

FIGURE 21.1. The location of embedded IRAS infrared sources in the core of the ρ Ophiuchi dust cloud. The solid contour shows the boundary of the strongest molecular line emission from interstellar carbon monoxide (CO), the tracer of high density molecular hydrogen. Approximately half of these sources are believed to be T-Tauri stars; the other half are protostars. (R.B. Loren)

matter as they move through space, apparently escaping from a T-Tauri star. Only now is their relationship to star formation being recognized. They are "bullets" shot out from the T-Tauri stars and associated with molecular (mostly CO) outflows.

The protostar's surface is at about the same temperature as the sun's (5,000K), yet this is not yet a star. Its energy comes from gravity, which pulls matter inward and causes the surface to heat up in collisions with the infalling material. The sun was about ten times brighter than it is now when it was at this stage in its evolution.

Charles Lada and his colleagues at the University of Arizona have observed the actual infall of material onto a protostar in the ρ Ophiuchi cloud which has the poetic name of IRAS 16293-2422. They used radio telescopes to measure the Doppler shift of cold molecular material and also saw the outflow of hot gas in jets away from this object (see below).

Bruce Wilking and his colleagues, Richard Schwartz and James Blackwell at the University of Missouri-St. Louis, have searched for objects in the ρ Ophiuchi cloud which showed an excess of Hα emission (an optical spectral line produced by hot hydrogen) indicative of circumstellar gas. In regions of star formation, that is, the very clouds whose ominous darkness so fascinated Edward Barnard a century ago, this emission almost certainly indicates the presence of T-Tauri stars. Twenty-seven of the sources they found were coincident with infrared stars seen by IRAS, further evidence that these objects are dust-enshrouded protostars.

Wilking's newly identified T-Tauri stars in the ρ Ophiuchi cloud have been located on a Palomar Sky Survey image for this region (Figure 21.2). The T-Tauri stars which are barely visible, are indicated by the small lines with numbers referring to the catalog made by the original researchers. It is fascinating to compare both Figures 21.1 and 21.2 with the photographs of the same region taken by Barnard (Figures 3.2 and 4.3) to see just how far astronomers have come in technology and understanding since Barnard's time. Clearly Barnard could never have known that this "hole in the heavens" was a stellar nursery of extraordinary activity. This is the same dark region at which Richard S. Tucker once pointed his telescope and thought that the weather had suddenly become overcast because he could no longer see any stars in this direction (chapter 5).

The ρ Ophiuchi cloud contains enormous numbers of faint images which are T-Tauri stars, objects about to burst into full-fledged existence as honorable stars.

Jets Associated with Star Formation

The motion of the molecules in the dust clouds in the immediate vicinity of the protostars accreting matter show not just infall but outflow, in two directions. These are called bipolar flows—narrow, high-speed jets of matter that stream away for the incipient star. As they push into surrounding space, matter piles up ahead of them to form the HH objects, which move at speeds of several 100 km/s.

Figure 21.3 shows one side of a bipolar outflow as a corkscrew-like jet of emission emerging from the dark globule photographed by Bart Bok who pioneered their study. This object is so stunning that is worth special attention.

This southern hemisphere object was carefully examined by Richard D. Schwartz of the University of Missouri in St Louis. A small jet of luminous matter emerges from a dark globule which lies within the boundary of a very old supernova remnant known as the Gum nebula. The two dark clouds in Figure 21.3 are between 75 and 300 pc distant.

Figure 21.4 shows in schematic form what can be seen in Figure 21.3. A corkscrew-like jet emerges for the dark cloud. Inside, a hidden, spinning protostar is surrounded by an accretion disk. The rotation imparts a

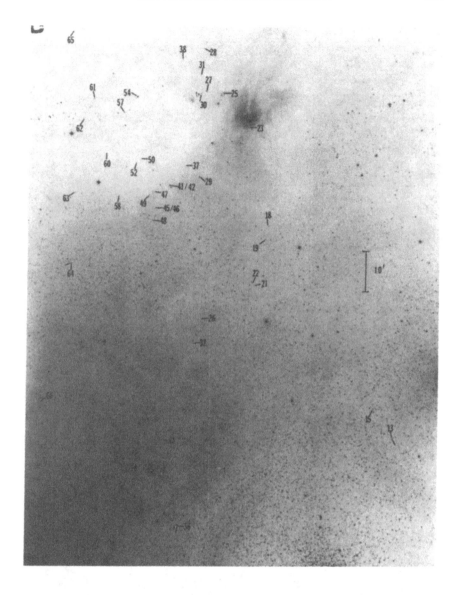

FIGURE 21.2. A highly blown-up negative print of a section of the ρ Ophiuchi dark cloud showing the location of several faint objects rich in hydrogen-alpha emission, indicating that they may be sites of current star formation. This photograph may be compared with Figure 21.1 and shows that the optically visible objects lie along the boundary of the dust cloud where some of their light manages to filter out. Their presence in the infrared data is more obvious, and in the near future infrared imaging should reveal the interior of this cloud in great detail. These two images show how far modern astronomers have gone since Barnard first photographed this dark nebula (Figures 3.2 and 4.3). (Bruce A. Wilking and the *Astrophysical Journal*)

corkscrew motion to the emerging jet, as will be described further. Some of the accreting material that smashes onto the protostellar surface escapes to create the bipolar outflow.

At the end of the visible jet is a faint nebulosity, an HH object. On the far side of the dark cloud another HH object can be seen, but the jet is hidden by the dust. These two emission nebulosities are called HH 46 and 47.[4]

Figure 21.3 therefore reveals dark clouds of interstellar matter, filled with dust and molecules, within which an incubating star signals that it is about to be born by propelling jets of matter into space. One of these pokes through the placental material on our side of the dust cloud. Surrounding that star, as implied by the presence of the jets, is a disk of material which will form planets once the star settles into a steady, mature life. And on those planets the rain of molecules from the surrounding dust cloud will form the matrix for the origin of life, but only if the planet is suitable for it to take hold. Then, someday, about five billion years from now, intelligent creatures on that planet will look out and discover interstellar matter and see in the distance a planetary nebula, a ring blown out from a star whose orbiting civilizations once called it Sol.

Another example of an HH object, perhaps a more obvious one, is shown in Figure 21.5. This is NGC 2261, also known as Hubble's variable nebula. It is focussed on a star called R Monocerotis (R Mon). The notation implies that it was long known to be variable. In fact both the stellar object and the nebula vary in brightness from year to year, but in no regular manner. This image was interpreted by Karen Strom[5] and her colleagues at the University of Massachusetts. Combined with the knowledge that R Mon is also a bright IRAS source, it appears that here a luminous young star, surrounded by a circumstellar disk, is also ejecting material in a bipolar flow. Although the visible nebula is not suggestive of symmetry, observations of the CO molecule show both a red- and a blue-shifted side to the flow.

FIGURE 21.3. Two ominous looking, dark markings among the stars. These are southern hemisphere examples of Bok globules, small scale versions of the dark markings that puzzled Barnard and other astronomers for so long. These two ghosts in space are dark, elusive, protective of what is happening within them. They are silhouetted against 4,000 stars in this image of a tiny section in the Milky Way, their edges faintly glowing from the starlight they reflect. In the larger of the two globules an extraordinary spectacle, a corkscrewing luminous jet, ushers the viewer in for a closer look. This is a jet of matter ejected from the core of a disk that has formed around a star about to be born. Faintly luminous matter is seen to glow beyond the end of the jet. This photograph was taken by Bart Bok on Valentine's night, 1978, with the 4-meter telescope of the Cerro-Tololo Inter American Observatory. (CTIO and NOAO via Astronomical Society of the Pacific)

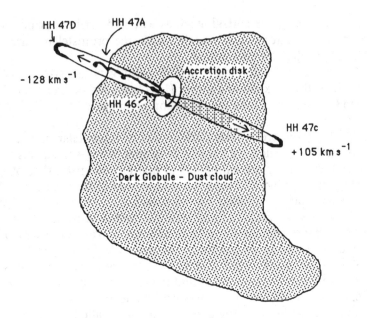

FIGURE 21.4. A schematic representation of the invisible goings-on within the dark globule shown in Figure 21.2. A rotating accretion disk has formed about a protostar and two jets of matter are explosively ejected along the axis of the disk. Herbig-Haro objects, faintly luminous nebulosities, are created at the shock front where the jets push up against surrounding interstellar matter. (Based on research by Dopita et al., note 4)

The explanation of bipolar flows initially involves the formation of an accretion disk around the incipient star. This disk (Figure 21.4) consists of matter drawn from the surrounding gas and dust cloud, which first gathers in a rotating disk-shaped volume of space before falling into the protostar.[6] The bipolar jets then emerge from the center of this disk.

What Drives the Jets?

The jet energy can be estimated from the molecular line observations which allow the velocity of the material as well as the mass to be calculated. The odd thing is that not enough energy is created by the relatively cool protostar to drive the jets outward as fast as they are observed to flow (up to several hundred km/s). The stellar energy is known because it is related to how much radiation the star emits, in turn derived from observations of its brightness and a knowledge of its distance. This gives the stellar luminosity, that is, how much energy the star actually emits. It turns out that in all cases of bipolar flows the jet energy is unrelated

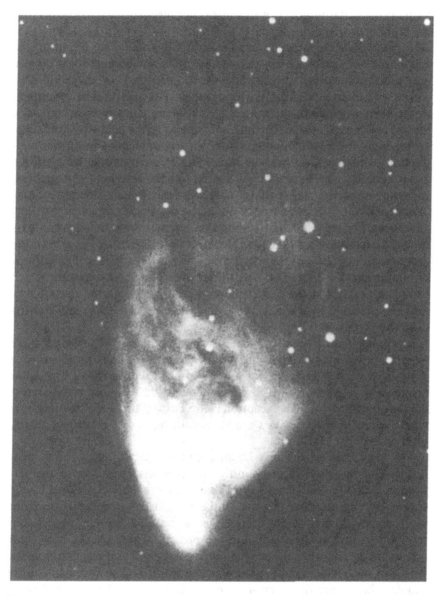

FIGURE 21.5. The Herbig-Haro object NGC 2261 also known as Hubble's variable nebula which emerges for the star R Mon. The swirling nebulosity has been ejected for the accretion disk surrounding a very young, barely formed, star. Light is reflected from the dust that is left over after the star formed. (Copyright Anglo-Australian Telescope Board.)

to the protostellar luminosity so some other mechanism must be driving the bipolar flow.

If magnetic fields were present in the cloud from which the protostar condensed, they would be drawn in with the gas; and because the protostar inevitably rotates (to conserve angular momentum) the magnetic field will be wound up something like a corkscrew. The field, in turn, is coupled to the matter—glued to the material, as it were. Later in the protostar's life the field has to escape (become uncoupled) or the field strength builds up so much from being tightly wound that it prevents any further contraction of matter onto the star itself. This is a continuing problem for astrophysicists who have not yet figured out the details of this process. In any event, some matter that is suddenly heated upon striking the star blasts back into space and rushes out along the field lines which, by now, have been twisted into tight spirals that emerge along the rotation axis in two directions. This structure has been observed as was shown in the photograph in Figure 21.3 and schematically in Figure 21.4. The outflow of matter along the jets also acts to remove angular momentum, which causes the matter in the disk to flow more rapidly into the central star; thus the presence of a bipolar flow may indicate that gas is still flowing onto the protostar.

Low-mass stars in the process of forming, including T-Tauri stars, have been found in many of Barnard's dark clouds, but stars more massive than eight solar masses do not appear to exist within them. The latter are mostly located in GMCs, which tend to be invisible behind great depths of distant dust. An exception is the Trapezium in the Orion nebula where seven massive O stars are located in front of a relatively nearby (400 pc distant) GMC.

The formation of jets may be an inevitable part of early stellar evolution, and the jets will profoundly influence the chemistry of the surrounding clouds. The chemistry in the jets is probably controlled by the shocks. A characteristic feature is the (2-μm) infrared radiation from molecular hydrogen at the outer edge of the flow. Here strong shocks required to produce the radiation are present. These shocks help form certain molecules while destroying others. They will also tear grains apart. The flow is supersonic and sweeps gas ahead of it. The CO maps of the bipolar show only the swept-up mass. The wind material is not seen, because it may be neutral atomic gas.

Water masers are also seen in star-forming regions, either in the protostellar atmosphere or in the immediate vicinity of the star. Densities in these regions are enormous by astronomical standards, up to 10^{10} cm^{-3}. Very-long-baseline-interferometry observations with high resolution shows that the stellar masers are all moving away from the star, an effect that can be explained by continuous ejection of material over a thousand years. The time estimate, as is the case for so many of the numbers derived in regard to star formation processes, comes from applying the principle of balance between forces, in this case involving a time dependence.

The types of stars that produce the bipolar jets range from the normal, solar-sized objects like the T-Tauri stars to much more massive protostars. They are all in condition just prior to lighting the nuclear furnaces at their heart, which will fuel the star for millions or billions of years to come.

When the star does begin to shine by its own nuclear processes a new wind is created, one that blows outward in all directions creating the stellar wind, a wholly different phenomenon. Such winds inevitably destroy the surrounding cocoon of dust and gas. In the early phases of the sun's life it, too, created a strong wind which, if it had kept up, would have evaporated the sun in a million years. Today the merest hint of a breeze, the solar wind, blows out from the sun and no longer removes significant amounts of matter, although it does touch our planet in significant ways.

A Summary of Star Formation

Diffuse clouds of interstellar matter gather to form dense structures. Energy originally derived from galactic rotation drives the clouds hither and thither. Shock waves created in supersonic turbulence causes matter to pile up in substructures, creating dense cores. Gravity begins to exert control and when the matter is decoupled from the cascade of turbulent energy—as the result of the formation of a density "wall" at the edge of the clouds—the material proceeds to shrink under the influence of gravity. Heat generated in this manner is radiated away as infrared radiation. The cloud thus remains cool enough so that gravity continues to shrink the object. A central condensation—the protostar—is formed, surrounded by a disk of material in the process of being accreted from the surrounding dust and gas cloud. Some of this gas finds itself in the twisted magnetic field region and rushes out into space to form a bipolar flow. The flow pushes up against surrounding interstellar matter, where piled up and heated gas glows and is seen from earth as a Herbig—Haro object.

Meanwhile, back at the protostar, the collapse under the influence of gravity speeds up. More heat is generated, and the surrounding dust is heated and forms an infrared source seen by IRAS and other infrared telescopes.

The protostar shrinks further, and just before the nuclear furnace turns on, the surface of the object convulses in brightness as matter falls in. The T-Tauri phase, also known as a pre-main sequence star, will evolve into a full-fledged (main-sequence) star when its core temperature reaches about 15 million K. Then hydrogen nuclei (protons) fuse with other hydrogen nuclei to form helium, and this process liberates nuclear energy.

This is the fusion reaction, the Holy Grail for physicists seeking abundant supplies of cheap power. It is also the root of the looming threat that hangs over humanity in the form of hydrogen bombs. It is sobering to note that in order to achieve a controlled fusion reaction nature requires stellar masses. This makes one wonder whether there isn't something intuitively odd about scientists working diligently in the laboratory to control fusion, for which nature requires gravity and a stellar mass. After all, if nature knew of a better way, wouldn't it be observable somewhere? Can scientists hope to do better in the laboratory? One wonders whether the only way to emulate nuclear fusion in a controlled environment is to build another star!

In the protostar the resulting nuclear explosion is kept under control by the gravity of the all the matter involved. Only at the very core is the nuclear burning occurring and the heat so liberated percolates to the surface and causes the star to shine with its own light. Gravity and internal heat now reach detente and the star neither shrinks any further nor expands. Some stars may pass through a temporary, variable phase as they seek this condition of stability.

Once the stellar furnace is lit, the star shudders, passes through a variable phase where its surface alternately expands and contracts (as does a Cepheid variable), and then settles into a condition of perfect balance. Now the force of gravity precisely equals the force produced within the star in the conversion of vast quantities of hydrogen into helium. In the sun the nuclear fusion process uses up about one million tons of hydrogen every second, yet the sun is so massive that it can happily burn the stuff for 10 billion years with the difference being negligible.

The Perpetual Cycle

Our adult star has now become a regular member of the galactic assemblage of stars. The vast majority of these will live to a ripe old age before dying gently, perhaps passing through a planetary nebula phase (Figure 15.1) and ending up as a white dwarf, ultimately cooling to invisibility. More massive stars will end their lives catastrophically and somewhat prematurely as supernovae.

It is possible that large numbers of dead stars, those that long since ran out of fuel, but did not explode in the process, exist in space, forever dark and lost from view. Such stars act as a drain on interstellar matter because whatever gas they used is removed from the cycle that carries gas into stars, processes it, and injects it back into space in explosive events. The recycled gas is continually laced with heavier elements that were formed in the nuclear furnaces of the stars or in supernova explosions, and so the chemical content of interstellar matter has inexorably changed with time. It once consisted of only hydrogen and helium but now, as the result of the birth and death of billions of stars since the galaxy's formation, interstellar matter is a cauldron of more complex atoms, many of them linked into molecules needed to form life on planets that orbit newly formed stars.

Possibly most of the mass of the galaxy will ultimately go into low-mass stars that will last longer than the age of universe. Star formation will then cease, but the Galaxy will become a tired hulk of faintly radiating stars that will, for a very long time to come, drift about in a true void because the interstellar matter that so rejuvenates the cycle will have disappeared.

How far we have travelled from Barnard's lonely nights at the telescope a century ago, when he patiently tracked a star for six hours while his

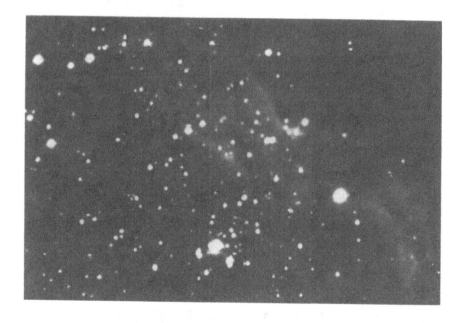

FIGURE 21.6. A recently obtained infrared (2.2 μm) image of the interior of the dark cloud seen in front of the M17 nebula (Figure 19.1). This is one of the most active star formation sites in the galaxy. The infrared image, a mosaic of more than 100 individual frames obtained with the infrared array camera of the National Optical Astronomy Observatories, reveals a cluster of young, hot O and B stars which had heretofore been hidden by the dust. More than 1,000 infrared sources have been found in this cluster. The streaks of emission at the right are ionization fronts between the molecular cloud and the emission nebula, M17. (Charles Lada, Steward Observatory, from a project with Ian Gatley and Daren Depoy)

FIGURE 21.7. An infrared image of the nebula, NGC 2024, located in north of the Orion nebula. The image was obtained with the NOAO Infrared Array Camera. Sixty-four 1 × 1 arcminute frames were pieced together to make this picture. The frame is 8 × 8 arcminutes on a side. A cluster of young stars never seen before because it was hidden behind dense a dust cloud (which contains CO, see Figure 20.1) was discovered in this image. (Elizabeth Lada, University of Texas, from a project with Daren Depoy, Neil Evans, and Ian Gately)

photographic plate hung exposed to the faint light from space. How far in imagination, too, we may yet go as we bear witness to the existence of the very seeds of life in those dark clouds whose image he so carefully caught.

Seeing in the Dark

At the time this book was being written a dramatic new technology was being developed which allows astronomers to at last see into the dark clouds. For years infrared telescopes had been used to probe these clouds, revealing young stars in the process of formation, previously hidden by the veils of dust. Now, new technologies which stand ready to reveal all that has so long remained hidden involve the use of imaging devices which allow infrared "photographs" to be made of the objects that lie hidden within or behind the dark clouds.

We end this chapter showing one such image. Figure 21.6 shows the infrared view of the nebula M17 whose optical photograph, replete with interstellar dust, is shown in Figure 19.1. Comparison of the two shows that the infrared image has laid bare to our view a cluster of young hot stars formerly hidden from view.

Figure 21.7 shows an infrared image of the Orion region, about 5° north of the Orion nebula. It exposes a stunning view on what happens inside and behind the dark clouds which kept Barnard and other astronomers arguing about "holes in the heavens" for so long. These two images have kindly been made available by the astronomers concerned; at the time of writing the interpretation of what has been seen in images such as this had barely begun. What is certain is that the infrared imaging technology is about to lead to a dramatic breakthrough in our understanding of star formation and the nature of interstellar matter, because now nothing in space remains hidden to human "eyes."

Notes

[1] O and B stars are the largest, youngest and hottest stars observed and they are found to be distributed along spiral arms of galaxies. Older stars are found to be more widely distributed, both in an outside the arms.

[2] Turner, B.E. (1988) "Molecules as probes of the interstellar medium and of star formation." *Galactic and Extragalactic Radio Astronomy*, New York: Springer-Verlag, p. 154.

[3] Larson, R.B. (1987) "Star formation, luminous stars and dark matter." *American Scientist*, 75: 377. This provides a fine summary of the topic.

[4] Dopita, M.A., Schwartz, R.D., Evans, I. (1982) "Herbig-Haro objects 46 and 47: Evidence for bipolar ejection from a young star." *Astrophys. J. Letts.*, 263: L. 73.

[5] Strom, K.M. et al. (1986) "Optical Manifestations of mass outflows from young stars: An atlas of CCD images of Herbig-Haro objects." *Astrophys. J. Suppl.*, 62: 39.

[6] Accretion disks exist in a wide range of astronomical situations and are discussed in more detail in *The invisible universe revealed* by Gerrit L. Verschuur. New York: Springer-Verlag: 1987

CHAPTER 22

A Mysterious Interstellar Matter[1]

In the Right Place at the Right Time

At the E.O. Hulbert Center for Space Research of the Naval Research Laboratory (NRL) in Washington, D.C., a group of astronomers, likely never to man a battle-cruiser or set foot on foreign beachheads, enjoy Navy support to engage in research on such unlikely topics as quasars. On this day the walls of a corridor are hung with yard upon yard of computer output showing the strength of sixteen radio sources, distant quasars, that have been observed five times a day for seven years.

The scene is set for a great discovery, but Ralph Fiedler, newly hired post-doc from the University of Iowa, did not know that as he placed a chair in the corridor and leaned back to get an overview of all the data. Then he noticed a subtle problem. The intensities of fifteen sources had been calibrated with respect to the quasar 3C286, believed to rock steady, but the charts showed the brightness of all the sources changing together, even if only by a few percent, a clear sign that systematic problems were affecting the analysis.

As he meditated on what next to do he spotted a glaring four-month gap in the record for quasar 0954+658.[2] He asked around and learned that previous workers had thrown out those data, because the source had exhibited such extraordinary behavior that something had to be wrong with the measurements. Fiedler made a mental note to check later and see for himself.

His first task, though, was obvious. He had to systematically recalibrate all seven years of data, which required writing a computer program to do the job. As the work proceeded he took time to briefly look over the raw data for 0954+658. He noticed that its variations were odd and his curiosity was piqued. The new plan was to take the information on several sources as a new standard. A few months later the program was ready and one night all the observations were reprocessed.

The next day he began to repaper the walls and spread out into a second corridor. Again he pulled out a chair and this time was pleased with what he saw. The joint meanderings of the sources were no longer visible, which meant that the calibration could be relied upon. His next step was also clear. He had to carefully examine the sources, one after another, and identify intrinsic source variability and scintillations. The analysis could now begin in earnest. He was not yet aware that missing data for 0954+658 had automatically been processed and were waiting for him!

Keeping Time with Quasars

The observations had come from the radio interferometer at the National Radio Astronomy Observatory in Green Bank, West Virginia, and represented the output of a long-range collaborative program undertaken by the U.S. Naval Observatory (USNO), the nation's time keepers, and astronomers at the NRL. The interferometer, consisting of two 85-foot telescopes, a 45-foot some 30 km away, and a 46-foot telescope approximately 35 km distant, but at right angles to the line of the other three, had for more than a decade done sterling work in the study of radio source structure. In the 1970s, Sir Martin Ryle and Bruce Elsmore at Cambridge, as well as Ken Johnston at the NRL with Campbell Wade at the NRAO, demonstrated that interferometric observations of quasars could give information about the Earth's motion in space. This motion is complex and requires the likes of a Naval Observatory, or a Greenwich Observatory in Britain, to keep our communal clocks running accurately.

In 1975 the USNO and NRL began to discuss how they might incorporate the Green Bank interferometer into their research. The Time Service wanted to use observations of quasars to supplement optical timing of stellar transits and other techniques for measuring time. The NRL was interested in quasar variability.

Quasars are powerful sources of radio emission situated billions of light years from earth. Many have very small angular diameters, some only fractions of a thousandth of a second of arc across. A radio interferometer measures the angular location of radio sources very accurately and can also measure the precise time when the quasar crosses the meridian each day. This was just what the timekeepers at the USNO wanted to supplement other efforts to measure time.

Repeated observations of the quasars also give the location of the interferometer telescopes tremendously accurately. The potential in these observations, as the astronomers in 1973 had pointed out, was so great that after repeated observations the location of the telescopes could be pinpointed to the nearest centimeter. This meant that motions of the Earth's continental plates could be observed, provided one telescope sits on one plate and the other on an adjacent one. But in Green Bank all the telescopes sit on the same "stable" continental plate, so the USNO wanted to find out exactly how that platform, and hence the Earth beneath it, moved in space. This would allow them to determine the length of the day to accuracies which the astronomical timekeepers had struggled for centuries to achieve.

In June 1977 word reached the potential collaborators that the Green Bank telescope was to be taken out of service. This disaster could only be avoided if they took over its operation, but that required money and the USNO's three-year budget was about to be submitted. How could they, at such short notice, include a new line item: one used interferometer?

"We scrambled like mad," Gart Westerhout, Scientific Director of the USNO, confesses, "and we succeeded." In October 1978, with the blessings of the National Science Foundation, the use of the interferometer was handed over to the USNO.

Five times every day sixteen quasars were observed and interference fringes recorded. An interference fringe contains two pieces of information: the timing of the fringe, known as phase, and the amplitude of the signal. If you perform the appropriate computer hocus-pocus the timing gives the source position in the sky and, as every science student knows, repeated measurements lead to greater accuracy.

The amplitude of the fringes gives the strength of the radio emission from the source. Ever since the mid-1960s it was known that radio sources vary with time. Some of the variability is due to events in the nucleus of the quasar—explosions that eject vast quantities of matter into space. Quasars also twinkle, or scintillate, due to irregularities in ionized interstellar matter between the source and the earth. (This is similar to the phenomenon that causes stars to twinkle as their light passes through the atmosphere.)

The collaborators planned to share the data. The USNO group would use the timing (phase) information. They did not care how bright the source was, as long as they could see it. The NRL astronomers, on the other hand, would pick through the quasar brightness data to learn more about intrinsic source variability and the effect of interstellar scintillation and a second here or there did not concern them too much.

Daily observations were made at two frequencies (8,085 and 2,695 MHz) which would particularly help distinguish the difference between intrinsic variability, manifested first at the higher frequency and months later at the lower frequency; and scintillation, which occurs on 10- to 80- day timescales and is more pronounced at lower frequencies. Every 30 seconds information was stored in the computer at the control center in Green Bank. Averages of 10 minutes of phase and amplitude for each source were then obtained. This was a lot of data, because there were four telescopes at two frequencies, which meant twelve possible pairs, at each of two polarizations.

At the Earth's Orientation Parameters Division of the Time Service of the USNO, David Florkowski and Dmitrios Matsakis and their collaborators "massaged" one copy of the daily dose to obtain the phase information. This was passed to the official keepers of the clock who combined the quasar data with satellite observations, lunar ranging measurements, and star transits to set time. They, in turn, provide tables, accessible by modem or by mail, which allow mariners to cross oceans safely, satellites to be parked in orbit, complex secret codes to be generated, and businessmen to meet for lunch to the nearest microsecond if they so desire.

The NRL group, led by Ken Johnston, took their copy of the data tapes and began to look at the source brightness information in three-month intervals. A number of post-docs worked on the project from time to time, always looking at small segments of the data. In 1983 a preliminary report was published. But all the data had not yet been compared. In October 1985 Ralph Fiedler arrived and was given the task of making sense of all seven years's worth of information. He took up station in the corridors in order to get an overall perspective of the data.

Accurate measurements of the radio source intensities are difficult because many unpleasant things conspire to confuse the task: interference from electrical fences, cars, satellites, breakdowns in receivers, computer malfunctions, and electrical leakage in cables are all capable of causing spurious results. With enough patience each problem can be recognized and taken care of, by making repairs, correcting for it, or, as is sometimes the case, disgustedly throwing out the data. When in early 1981 the source 0954+658 showed four months of extraordinary variability it was initially vetoed. Fortunately, the baby wasn't thrown out with the bath water and the raw data tapes had remained untouched in the archives.

A Peak Experience

After the walls were repapered with the newly calibrated data it took Fiedler a week to reach 0954+658. Then he saw why it had been thrown out. Its signal indeed varied in an extraordinary manner, like nothing ever seen before (Figure 22.1). At 2,695 MHz the source brightness rose for a week or so and then faded to half the initial brightness and stayed low for nearly two months. Then the signal rose again, increased above its normal level for a week, then slowly returned to normal. The whole scenario took nearly four months. At the higher frequency the source brightness rose and fell four times, on one occasion more than doubling in brightness in a few days. This behavior was unprecedented, and as Fiedler stared at the records he began to experience that moment a scientist lives for: the thrill of seeing something for the first time, a new manifestation that no one else has ever seen, anywhere, at any time! He knew that these were not spurious meanderings of the receiver, but a real, hitherto unseen, phenomenon.

"I was elated," he says, still excited by the memory. "I ran shouting and screaming down the corridors." Understandably so. This was Fiedler's peak experience (chapter 12). He was elated for weeks. "It is like a biker's or runner's high," he volunteers, "but this lasted much longer."

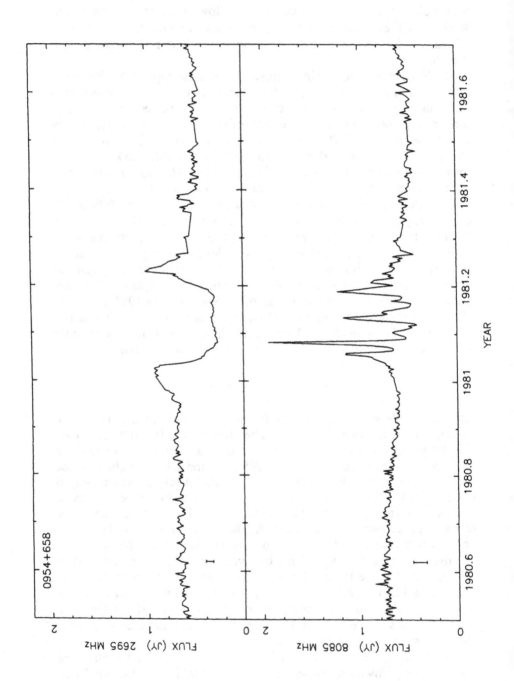

After the elation of a peak experience subsides the scientist must confront the task of effectively communicating to his colleagues what it is he discovered. Thus begins the next adventure, the interpretation of the data.

Fiedler's excitement drew a crowd that stared skeptically at the charts. The signals from 0954+658 looked interesting but were too incredible to be real. Something had to be wrong. No source could do that! They wanted more proof. Was the signal seen on each of the baselines between the four telescopes? Yes, it was. Was it in the raw data? Yes, it could be seen even there.

"So what?" the skeptics asked. "Perhaps it's due to interstellar scintillation." No, because scintillation is weaker at higher frequencies and the variations of 0954+658 were stronger at the high frequency. Also, scintillations produce only a few percent brightness changes over weeks to months. This source more than doubled its brightness and thus could not be produced by scintillation.

Was the variability intrinsic to the source? No, because intrinsic variability of 0954+658 had a clear signature also seen in the data; slow changes at the high frequency were followed by similar variations at the lower frequency fifty days later. This new phenomenon occurred at the same time at both frequencies.

There was one way out. The new phenomenon could be due to something that had drifted between the source and earth. This had caused an apparent occultation (the disappearance of the radio source) at one frequency and strong variability at the other. But what sort of structure could do that?[3]

Fiedler searched all the wallpaper and found two more sources that showed the odd behavior at the low frequency, even though nothing of significance was seen at the other frequency. Nevertheless, these two cases represented the confirmation he needed, and even the skeptics were impressed. The stage was set for the next phase, the tough job of finding an explanation.

Good science requires formulating theories to account for new observations; otherwise the words "So what?" will forever resound down the corridors. At this point Brian Dennison from Virginia Polytechnic Institute and State University returned to the NRL on a one-year reassignment. He was an expert on interstellar scintillation and was imme-

FIGURE 22.1. The signature of the extreme scattering event. The radio brightness variations of quasar 0954+658 at the two frequencies are shown. At 2,695 MHz the source flux rises then falls to about half its normal value and then rises several weeks later before returning to normal. Extreme brightness changes are seen at 8,085 MHz while this is occurring, on one occasion more than doubling its flux in two weeks. (Ralph Fiedler, Naval Research Laboratory)

diately excited by the source behavior. He thought that some extreme form of radio scattering or scintillation due to ionized gas clouds between the source and earth was involved. Fiedler, on the other hand, suspected diffraction because the behavior of the signal at the higher frequency was not unlike that seen when the moon occults a radio source. Ken Johnston, who had been the project's guiding light from the start, now took the role of devil's advocate as the other two struggled with one idea after another. Finally, everyone was satisfied that a reasonable explanation had been found and they sent their report to the British scientific journal *Nature*, which of course sent it to the usual anonymous referees. One of them, Anthony Hewish of Cambridge University, who shared a Nobel Prize in physics, famous for his work on pulsars and interstellar scintillation, made himself known to the authors. He thought they were on the right track but suggested refinements to the theory which would make it work even better. They delightedly accepted his suggestions and invited him to join in publication.[4]

What Can It Be?

The extreme scattering event (ESE, as it is now called, although some astronomers refer to it as a Fiedler event) was triggered by a very small structure with a clearly defined boundary that drifted in front of 0954 + 658 in 1981 (shown schematically in Figure 22.2). The quasar is only 0.39 milliarcseconds in diameter, and the ESE caused major changes in source brightness in fifteen days. From data at both frequencies it was inferred that the mystery object had to be moving across the sky at the rate of 0.09 milliarcseconds per day. If the object were at the quasar distance this implied a velocity 500 times that of light, unreasonable indeed. Bringing it closer to home the researchers recognized that velocities of 200 km/s are possible in the halo of the Galaxy, in which case the object would be 1.3 kpc distant. It might actually be much closer, but they decided to work with this figure. To account for the two month minimum at 2,695 MHz, the object had to be seven astronomical units (AU) across, very small by astronomical standards (the AU being the Sun to Earth distance).

Inside this small volume at least four or five patches of ionized material must also exist. These inhomogeneities had to have a diameter of 0.5 AU and caused scattering of radio waves. The theory predicted that scattering at the low frequency would bend the radio waves outward rather than focussing them, as a lens might (see Figure 22.2). This meant that as the object moved in front of the quasar, and again when the quasar reappeared, an enhancement of the source brightness would be expected at the earth. The structure was essentially projecting a irregular image and as the earth moved through this image two maxima separated by a

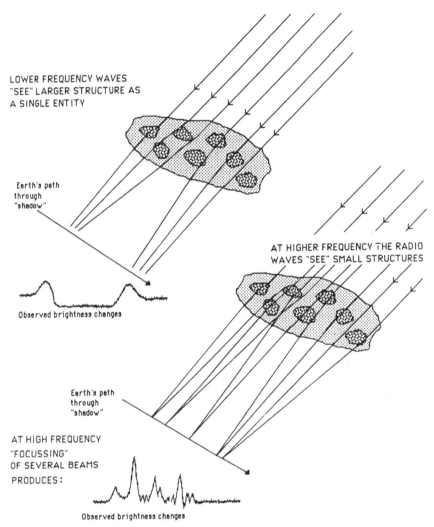

LOWER FREQUENCY WAVES
"SEE" LARGER STRUCTURE AS
A SINGLE ENTITY

Earth's path
through
"shadow"

Observed brightness changes

AT HIGHER FREQUENCY THE RADIO
WAVES "SEE" SMALL STRUCTURES

Earth's path
through
"shadow"

AT HIGH FREQUENCY
"FOCUSSING"
OF SEVERAL BEAMS
PRODUCES:

Observed brightness changes

FIGURE 22.2. A schematic view of the model to account for extreme scattering events. The low frequency radio waves see only the large structure, which deflects the radio wave paths so as to produce a shadowing effect with enhanced emission (concentration of waves) around the edge of the shadow. The higher-frequency waves see the inhomogeneities, so that several scattered beams are produced. In some directions they reinforce to produce enhanced radiation at the Earth.

deep minimum would be observed (Figure 22.2). The scattering effect at the higher frequency is different. Each of the inhomogeneities independently deflects the radio beam, so that a lot of separate beams emerge which may reinforce each other to produce bright spots. (This is also shown in the diagram and the computer simulation of the phenomenon.)

If the earth happens to be located with respect to the distant clouds so that it moves through these beams in the appropriate way (this will not always true, as is attested by the other two examples), a series of brightness increases would be seen at the higher frequency. The theory seemed to fit the observations very well.

Sixteen sources had been observed for seven years, and another twenty sources for two-and-a-half years, and a total of four clear events and three less dramatic ones had been spotted. Statistically this implied 100 of these structures per cubic parsec of space. This is an incredible thousand times greater than the number of stars, a conclusion which makes the ESEs one of the most startling astronomical discoveries of our age. Interstellar space may be populated with swarms of unseen objects more numerous than stars. This was an interstellar matter of the most enormous importance, a new constituent in the universe.

But what are these objects? How much matter do they contain? The mass of the structure could be calculated because in order to account for the scattering the ionized inhomogeneities required a density of 4,000 electrons per cubic centimeter. Yet the total amount of mass in all of the hundreds of billions of these objects is then a trivial 100 solar masses, completely insignificant compared to the 10^{11} solar masses in the Galaxy.

The most intriguing thing about the hypothesized structures, a point that the authors of the report hesitated to emphasize but did allude to, is that when when their gravitational stability was considered these objects could not exist. The equations of balance (chapter 14) did not work. The ESEs present the most extreme case yet uncovered of the "missing mass" phenomenon, that mystery which applies to so many structures in the universe that do not appear to contain enough mass to hold them together. What is not at issue, though, is that the the objects responsible for ESEs must be more common than stars.

Will We Ever Know?

Fiedler and his colleagues next planned to launch the largest monitoring program ever undertaken on a major telescope. Six months of daily observations of hundreds of radio sources were performed with the 300-foot telescope at the NRAO in Green Bank. It was hoped that when an event was spotted, other radio telescopes, each performing a different measurement, would quickly be brought to bear on the mystery of what it is that is passing by in the depths of space. In particular, very long baseline observations will reveal changes in source structure during the event. If the predictions are correct they will appear to be "sparkling" as patches of radio emission come and go around and in front of the source as the structure drifts by. At the time this book was completed the six-month observing program had netted three more events, one in progress as observations began, which was not recognized until too late, and another which was hampered by a telescope breakdown. Thus the mystery remained. However, the importance of the work had become so great that plans were launched to convert the interferometer so that it could make higher sensitivity observations of quasars so as to search for more of these events with hundreds of quasars included in the program every day. Hopefully, before too long, the mystery of these strange events will be solved.

Notes

[1] This chapter is reproduced with kind permission of *Astronomy Magazine*. A modifed version appeared under the title "Something passing in the night" in the November 1987 issue.

[2] A code indicating right ascension, in hours and minutes, and declination, in degrees, +65.8 in this case.

[3] It is interesting to speculate that the phenomenon might have been ignored, and possibly misunderstood, if data at only one frequency had been available.

[4] Feidler, R.L. et al. (1987) "Extreme scattering events caused by compact structures in the interstellar medium." *Nature* 326: 675.

CHAPTER 23

An Extraterrestrial Matter

A "Fateful Meeting"

One day in 1887 in the train station in Amsterdam a Dutch gentleman was sitting in the waiting room, having just missed his train back to Haarlem. As he whiled away the time he idly glanced up and spotted a young lady hurrying by the door. "That little coincidence determined the course of my life and I still curse the moment," he was to write years later.

Three years before that fateful day, this gentleman, Willem Marx, had a relationship with a young lady named Liza. She studied at the music school in Amsterdam and gave him piano lessons. Their friendship consisted of making music together and even creating an operetta. One day Liza brought along a friend, Marie Koning, and together the three went rowing on the Amstel, the great canal that runs through Amsterdam. They had fun, drank coffee at the Dremenbridge, swung on the swings, and laughed a lot. But later Liza began to cry because Willem seemed to be flirting too much with Marie. In deference to their friendship with Liza he decided he would make no effort to contact Marie after that day.

His relationship with Liza had long since ended by the time he missed his train. Thus it came as a pleasant surprise to recognize that the young lady hurrying past the waiting room was the same Marie whom he had met three years before. He got up and called after her. In the few moments they had before she caught her train they chatted and she told him she had moved to The Hague. Addresses were exchanged and then she boarded the train.

Willem wrote to Marie and they arranged to go on an outing. After that they were to meet regularly and go on romantic walks in the flower gardens of Keukenhof. In due course an engagement was followed by marriage and later they had a daughter. But, alas, Willem came to regret that fleeting glance at the railroad station because, it was a very unhappy marriage and ended in divorce.

The moral of the story is that if just 100 years ago this young Dutchman had not missed his train and, once stranded, had not happened to look out of the waiting room door at a particular instant, you, dear reader, would not be turning these pages. Willem and Marie were my great-grandparents.

Small incidents—random events—alter the course of history. A train missed, a wrong turn taken, or a head lifted at a fateful moment; such trivia lie in our past and alter destiny. The passing of such incidents is seldom noted and their consequences hardly dreamed. So it is also with matters of an interstellar nature.

We Are What We Are Because

As we come to the end of our explorations of interstellar matters, consider what we have learned. Let us do so from the context of where you are seated, now. Who are you? Where did you come from? And in answering, give our imagination license to explore in the context of what has been discovered.

The symbols on this and the preceding pages have, I trust, conveyed information, and may also have entertained. Consider, for a moment, the written word. What an extraordinary phenomenon! Members of one species on this isolated planet have developed a remarkable way to communicate. You are able to look at these symbols on a sheet of paper and can understand what is written. Your imagination may have been stimulated by the words and you may have had new thoughts and ideas that never occurred before.

Now consider this: You are able to read these words because of the way life on this one planet orbiting one star in one galaxy happens to have evolved—that is, has passed through a sequence of phases we label with a broad brush called evolution—*precisely* the way it did. If it had not been for the way *all* the pieces fitted together during the last five to fifteen billion years, you would not be here.

None of us would be here if the history of our planet, or of terrestrial life, had been only slightly different. The differences may seem insignificant from a local perspective, but they produced consequences that resonate through time. Evolution has a way of amplifying tiny factors whose consequences are inexorably stamped on the future. Obviously we can never be aware of how different life on Earth would have been if some small, hidden factor had been altered a billion years ago. In the case of the example above, I happen to know that if a certain gentleman had not missed his train in 1887 I would not be here. (He later wrote about that event and his letter survived.)

Collectively, we can only speculate on how different life would be if the asteroid suspected of causing the extinction of the dinosaurs had struck a million years earlier or a million years later, or had fallen in the oceans instead of on land. Even then, the messages passed through time by the geological record are not as detailed as a handwritten letter from an ancestor.

Consider also whether a particular star exploded to nudge an interstellar cloud closer to collapse five billion years ago. It, too, determined whether any of us would come into existence. Our destiny also required that specific lumps of interstellar matter of critical mass gathered to form a planet at the right distance from a star, a planet whose polar axis would be sufficiently tilted in its orbit to provide the seasons. This planet had to accumulate and hold enough water to feed a wondrous cycle of rain, erosion, and land formation that painted the continents with vegetation so that we might someday marvel at the magnificence of it all, even as we cut down the trees in order to make the paper upon which we print words to share our knowledge.

We can go further back. If it were not for the way ultraviolet radiation cooked up the suspended molecular ooze in icy mantles around interstellar dust grains these words would not have been written or read. If dust particles had not gathered together in grains to provide ideal conditions for holding water, and if comets had not rained down upon our earth, there would be no oceans and no vacation spots at which to enjoy tiny fractions of our fleeting lives. And so it goes.

Astronomers and other scientists have barely begun to make out the jigsaw puzzle of life, a puzzle whose pieces lie scattered through space and time. But because the human mind has an extraordinary urge to identify and place together such pieces, we believe that we are beginning to understand. One of the key elements of astronomical discoveries of the past decades concerns the sequence of events that befell earth, those that lead to its formation and sowed the seeds for life and which made us the way we are.

Can We Even Begin to Comprehend?

Fred Hoyle, gadfly astrophysicist, lurks in the Welsh hills thinking about the universe. He has made a career of disturbing the establishment and some of his words set the tone for our meanderings on these issues. Concerning our efforts to see beyond where we are at any stage in history, he wrote, "It seem[s] highly unlikely that the knowledge of the day could possibly be adequate to deal with the whole Universe: and yet if one did not proceed as if it was, it would be necessary to think *outside* what was already known, and how is it possible to think outside one's knowledge?"[1]

This challenge, to think outside what is already known, is a journey many a scientist must undertake if he wishes to cross the frontier of knowledge. But in order to communicate what is found, the return trip must also be made and then the report must be phrased in terms related to what is already understood about the nature of reality. The scientist cannot do otherwise. Sometimes the intrepid voyager may, for a while at least, sail out of sight of land, the land of agreed-upon doctrine, and there discover new worlds, new thoughts. The problem then shifts because a description of the unknown beyond the horizon may not be readily understood. Then the task of communicating insight may be made nearly impossible and history is replete with incidents where a scientist saw too far ahead and was no longer understood by his or her peers. Alfred Wegener's insight into the existence of continental drift was an example of this. Even Darwin saw patterns and understood the nature of evolution decades, even a century, before the majority of informed people understood.

Of necessity, the fruits of discovery must always be communicated in words the mainlanders, the establishment as it were, can understand. Thus science moves forward a tiny step at a time, because the system, the combined operation known as organized science, has to move along as well; and as is true of all large systems, inertia determines how rapidly ideas can be changed, invariably not fast enough to the great frustration of a seer.

"To think completely outside (of what is already known) is of course impossible," Hoyle reminds us. "Nevertheless, one can go beyond the detail of what is currently known, provided one maintains the "style" of physics, and indeed some of the most profitable adventures in science have come in just this way." Thus Sir Fred tempts us to let our imaginations run free and not fear the possibility that we may have thoughts that do not meet with widespread approval. Pursue them in the way scientists do, he advises, by testing ideas against experiment. But do not hesitate to unshackle your mind for the journey.

Another challenge Hoyle throws out is for us to examine the question: "Did the whole Universe come into being, all in a moment, about ten billion years ago?" Rather than quibbling about details and making calculations of the universe's age, he asks that we consider the possibility that the answer to this question is *no*. He wonders whether "the attribution of a definite age to the universe, whatever it might be, is to exalt the concept of time above the Universe, and since the Universe is everything this is crackpot in itself."[2]

Such a provocative idea causes astronomers to disagree, for they dare not move too far from what the data show. It is indeed difficult to avoid the conclusion that the redshift of galaxies and the existence of a cosmic microwave background are anything but evidence for an expanding universe and hence a start to time. But Hoyle is suspicious. "I would regard the need for the Universe to take precedence over time as a knockout argument in favour of a negative answer to the above question." Who, indeed, are we as a species to dare ask such mighty questions as concern the origin of the universe and in unique arrogance believe we may have the correct answer within cosmic seconds of the asking. After all, from a cosmic perspective we, as conscious beings, have only just arrived on Earth, yet we believe we already have the answers at our fingertips!

The Origin of Life

We have learned that giant molecular clouds in space contain the molecules necessary for life, and that photolysis in icy layers around interstellar grains is capable of making amino acids and other molecules, which exobiologists once needed to account for the origin of life in the Earth's primordial environment.

We have learned that water from a million (or perhaps ten million) medium-sized comets is enough to fill the oceans. (Perhaps water from other comets also washed the shores of Mars aeons ago.) But it is those same comets, as we learn when we look more closely, that are polluted by the organic substances which are the stuff of life. What happens when more of those rain on Earth?

Darwin once pondered whether "in some warm little pond, with all sorts of ammonia and phosphoric salts, light, heat, electricity, etc., present . . . a protein compound was chemically formed, ready to undergo still more complex changes." This, he suggested, might form molecules which could reproduce and begin the process called life.

With the above thoughts in mind, and knowing something about interstellar matters, let us "boldly go where no one has gone before"[3] and bear in mind another of Hoyle's questions: Is it possible that the basic mysteries of the universe may be answered by a science other than astronomy? "To most astronomers the thought of information crucial to cosmology being derived from biology will, I suppose, appear ludicrous. but the universe is *everything* , and to omit information from any source, especially biology with its vast storehouse of information, would be truly ludicrous."[4]

"There are those who are so uncomfortable with new situations that their practise, on hearing a new idea, is to search for an immediately overriding objection to it. I work in the opposite way," Hoyle says. "To begin with, I search for good things to be said about a new idea. If some emerge, and especially if they look strong, I then turn to criticism." As will we.

Hoyle and his collaborator, Chandra Wickramasinghe, have been stirring up a tempest by suggesting not only that the molecules necessary for life exist in interstellar space, but that life itself resides there, albeit of a very elementary sort—bacteria, for example. In defense of this claim they point out that laboratory samples consisting of dead bacteria can simulate the extinction spectrum of starlight (chapter 10). Their critics do not accept this, because dead bacteria (*e. coli,* for example) produce far more structure in the extinction spectra than is observed in space. The critics do not quibble that the spectra prove the existence in interstellar space of the constituent atoms of which bacteria are made. The only argument is about how complex the combinations of such atoms may become. Clearly mixtures of carbon, oxygen, nitrogen and hydrogen exist in the dust clouds, but many permutations, not just the carcasses of bacteria, can produce the observed spectra.

These two intrepid scientists have gone further and suggest that comets, too, carry elementary life forms, viruses perhaps, which occasionally seed the planet and cause pandemic or epidemic havoc to our genes.[5]

Hoyle then asks, "Could this potentially very large and continuing source of organic material (comets) have formed the basis for the origin of life, rather than the comparatively trifling quantity of organics generated in terrestrial thunderstorms and other small-scale events?" It is from this perspective that we might let our imaginations wander, especially in this era when the human species, at least that segment exposed to the ramblings of a select band of astronomers, movie makers, and mystics, grows increasingly interested in the question of extraterrestrial intelligence. Let us broaden the question: Is there extraterrestrial *life* in galactic space, not just on planets?

A decade or two ago scientists endeavoring to answer the questions related to the existence of extraterrestrials had to come to grips with the question of the origin of life on earth. They evoked images of Darwinian tidal pools in which energy from the sun triggered chemical reactions; or perhaps lightning flashed through an atmosphere filled with methane and ammonia and nitrogen and water to trigger subtle chemical changes. These reactions have been elegantly simulated in the laboratory by the elegant experiments of Stanley Miller and Harold Urey (in 1953 at the University of Chicago), who circulated a mixture of gases believed to be representative of the early terrestrial atmosphere and sent sparks into the mixture. Amino acids and many of the molecules that have since been found in interstellar space were created, just as they are in the experiments of Greenberg, Donn, and Nuth (chapter 10).

Once the substances necessary for life existed in the oceans, exobiologists needed only to evoke a subtle process which would lead to the formation of cells, because life cannot be defined as existing until something inside a cell is clearly separated from its environment and, furthermore, the cell has the ability to replicate itself.

These were very unlikely events to expect in the oceans, as Hoyle reminds us, because even if those tidal pools did their job to concentrate the important molecules, their contents are infinitely diluted in the oceans. So how can we ever expect to find the molecules tightly concentrated, a requirement for them to develop greater complexity? What better place, in fact, than on the surface of interstellar grains, which exist in countless number in clouds many parsecs in diameter.

The stunning discovery that interstellar space is seeded with molecules of great complexity in extensive and pervasive dust clouds, which lie in great lanes threading their way within spiral galaxies (Figure 23.1), is there to remind us that we are free to ask whether there isn't another way to account for the origin of life.[6] At the very least we have a way

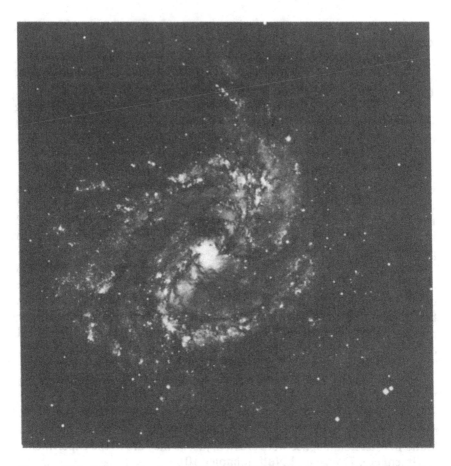

FIGURE 23.1. The spiral galaxy M83 (NGC 5236), which dramatically highlights the existence of vast dust lanes that thread their way between the stars in the spiral arms. Star formation is occurring within these dust lanes which contain the same mixture of gas, dust, and complex molecules found in Milky Way dust clouds (European Southern Observatory)

for ensuring that life emerges on earth without expecting that the first step, of making amino acids, for example, must occur in highly unlikely processes in the oceans or atmosphere. The discovery of the same molecules in space offers a vast reservoir of the stuff of life from the moment of a planet's formation. Their presence in space becomes a separate problem for astrochemists to solve. But their presence essentially allows us to look back in time to the epoch when the Sun and its planets formed from precisely the same material.

We know that meteorites carry amino acids and that all their building blocks exist in space. So why not explore this further, as Hoyle has done, and consider the next conceptual leap, that living things also exist out there. This, however, may be another idea whose time has not yet come, but that is no reason to shy away from considering the delightful implications that Hoyle's notion brings to mind.

Consider, for example, that an asteroid/cometary impact did hasten the extinction of the dinosaurs. Is it possible that it was not the climate change, but a viral visitation that came with the interstellar visitor that did the trick? Also, if we allow that the seeds of life, if not life itself, exist in the interstellar clouds, then our questions about life in the galaxy take on a new aura. After a planet is formed, provided only that it is at the right distance from its star and has sufficient mass (hence gravity) to hold the gases and water in place, the conditions for life may be as quickly established as they were on earth. We no longer have to be concerned about the tedious formation of the precursors of life in oceans or tidal pools or in the primitive atmosphere because they arrived with the package labelled "planetary formation."

If primitive life exists in space the most basic questions related to the origin of terrestrial life may become moot. We then have to worry only about the more complex phases, the formation of cells, for example, rather than proteins and DNA. This would raise the odds to near certainty that, whenever possible, life will evolve on all other remotely suitable planets. Whether it evolves to levels comparable to that on the Earth will depend on how closely similar to the earth such planets might be.

Sir Fred Hoyle reminds us of a few other things most scientists would rather not confront: "Could the vast storehouse of information necessary for the development of biology have been accumulated in only ten billion years?"[7] If you believe so, he says, consider what has happened in the last four billion years. The beginnings could not have occurred in the vast holocaust of radiation that existed on the early earth exposed to solar ultraviolet rays and its particle wind. Yet there has been virtually no change in the intricate biochemical complexity of life over the last three to four billion years. The enzymes in living systems are virtually unchanged from the cells of a human to the most primitive cells, similar to those believed to have existed on the early Earth. "Hence we a have a system without a promising beginning and with no change of the crucial aspects of the life system over the last one-third to one-half of the ten billion year time interval." His point is that life popped into existence at an unfavorable time and has nevertheless remained essentially unchanged ever since, a very unlikely scenario. Far better to explore whether life begins in space and survives a passage onto the surface of the incipient earth, well after the solar holocaust has died down, in that later epoch when comets may have visited in profusion and brought the water that filled the oceans as Greenberg and Chyba would have us believe. The stuff of life, and perhaps primitive life itself, may have arrived in those same comets.

Interstellar Origins

Hoyle and Wickramasinghe have looked at modern biology to find out how sure biologists were of the terrestrial origin of life and concluded that the facts were overwhelmingly against such an origin. "The terrestrial origin would require happenings every bit as miraculous as the views of the religious fundamentalists," they assert and so they turned to interstellar matter once again.

"We know from astronomical studies that the grains are mysteriously connected with a whole range of phenomena: the rate of condensation of stars; the mass function of stars; magnetic fields; spiral arms in galaxies; and quite probably the formation of planets. *Not one of these phenomena has been explained in better than fuzzy terms, just as the views of imaginary travellers in a spaceship would be fuzzy if they attempted to explain terrestrial fields, walls, and ditches as products of blind forces of nature.*"

We may not always agree with Hoyle, but we can explore his fascinating ideas to see where they may lead. Two decades ago no one would have believed there were dozens (probably hundreds) of species of organic molecules between the stars. It was back then that scientists still struggled to explain the formation of amino acids on the early earth by invoking lightning flashes and ultraviolet light illuminating a mixture of gases believed to exist here five billion years ago. That phase in the origin of life has clearly been shifted into space. Is it then so difficult to imagine that even more complex processes, as yet unseen, may also be occurring out there? It will take a lot more research to answer these questions, but that should not hamper the asking.

A serious implication of Hoyle's scenario is that if life in viral and bacterial forms exists in space, it may bring epidemics to our planet.[8] Extraterrestrial viruses, for example, may have been responsible for major evolutionary changes in terrestrial history. As support for this notion Hoyle has shown that the spread of flu epidemics, for example, are best understood by a broad injection of the virus, over a very wide area, rather than by assuming it emerged in one geographical area and was then carried outward from there. His conclusions, however, met with immediate and widespread criticism from the medical community. Nevertheless, they are food for thought.

Extraterrestrials?

But let us return to the suggestion that the building blocks of life, if not primitive life itself, exist in the interstellar dust clouds. What is the implication for life on other planets orbiting distant stars? This new model would imply that life will emerge as soon as it can on any remotely suitable planet. But how many of those life forms will ever be like us?

To answer this question we first take a broad view of life on earth and bear in mind that small incidents can have great consequences as time multiplies its effects. A glance toward a waiting room door alters time lines because without that look this book would not have been written. Similarly, life on earth is what it is because the moon causes tides and affects the evolution of life in the shallows around continents. Because continents drift and grind into each other to raise mountain chains, air masses alter their flow and redefine climate and weather. If the surface of the planet had not cracked and allowed lava to ooze out, continents would not drift and provide the delightful instability upon which evolution thrives. Thus bipeds capable of exercising their brains in order to survive would also not exist. This scenario, sketched with countless variations to which we must add catastrophic visitations from space, may be manifested on all planets. If so, we would then have to allow that life exists on millions, perhaps billions, of planets in the Galaxy, not to mention in other galaxies. But how many of those extraterrestrial species are available for us to talk to, by any means at all?

Again, consider the Earth. Look around at its myriad species. There are a million catalogued—estimates seem to vary every time I read another article on the subject—and some biologists report that the jungles of the Amazon may contain millions more. The stunning variety of life on earth today is the current expression of the Earth's evolutionary past, and the hundreds of millions of species that may already have gone extinct in ages past are barely a memory.[9] Had that past been any different, on whatever scale, had the continents not drifted or the land masses not gathered in the north, or asteroids and comets not struck when they did, the variety of life we now see would be very different.

Imagine that there have been, say, 500 million species on the Earth since life emerged. With how many of the others could we hold an intelligent conversation? The answer is obvious—not one. Or imagine it this way: If we were to choose a random moment in time and visit a random spot on the planet Earth, could we talk to any creatures around us? Of course not. We can talk only with *homo sapiens.* Would it be any different if we were to set down on a distant planet? One chance in 500 million is not high odds for expecting twin civilizations in even hundreds of millions of inhabited planets.

The proponents of an expensive search for extraterrestrial intelligence using radio techniques ignore this issue and would have us believe that if we could randomly set down on earth ten million years from now (the assumed lifetime of communicative civilizations), we would be able to talk to our evolved selves far in the future. This peculiar idea reflects a hidden assumption. It claims that after we twentieth century humans arrive on the scene evolution, and all that it implies, suddenly ceases, and that *homo sapiens* ten million years hence will be so close to the way we are today that we can sit down and chat with them!

It may be argued that if it hadn't been for our nasty warmongering ways in the past, perhaps other protohominids might have evolved in different directions and would be around for us to talk to. However, it appears that there is a biological niche for only one "intelligent," communicative, civilized species per planet surface. The question is whether a similar niche has been filled on other planets, and how long its inhabitants will last. Even if such niches are filled, why would such creatures be any more like humans than the millions of other species on our planet? There may be life on millions or even billions of planets, but it will be at least as different from us as we are from other terrestrial species.

I return to my original point. If that gentleman in 1887 had not missed his train, your bookshelf would not have space for a copy of *Interstellar Matters*, and you would not now be reading these words. Because he missed his train who knows whether a teenager reading this book may yet be inspired to become an astronomer, study interstellar matter, and make some profoundly important discovery, etc.

A Note About the Future

The galaxy may be teeming with life as alien to us as the grasshopper, the black widow spider, or those remarkable tube-like organisms growing next to volcanic vents on the ocean floors. The only way we will find out is to fly to the nearest stars, orbit their planets, and see for ourselves. It will be a long, tedious experiment, sending out remote vehicles that take centuries to get there, but it will be no less rewarding than spending money on radio searches, also likely to take centuries to be successful. In either case none of us will live to see an exciting conclusion to such ventures. But of this we can be sure: until the human species considers experiments whose rewards will come ten generations down the line, we will never know.

If, on the other hand, as so many would believe, there are advanced species capable of flying between the stars, they will make a prolonged visit to earth only when they are ready. In the meantime, it may be more fruitful for us as we search to uncover truths about nature to remember that extraterrestrial life in the Galaxy will be very alien and that it is in any event very far away. Until they arrive we will do better to focus on treating our own planet with greater care as we reach into space, undertaking such adventures as we can afford.

It may be very difficult for our generation to confront the fact that because of political and economic reality we may not soon be walking the Martian deserts. From a cosmic perspective, this hardly matters. Someday it will happen. In the meantime, though, we must consider that we have evolved far enough that projects taking longer than the working life of an ordinary citizen will soon become a necessity. Until now, space exploration has always been planned and carried out on time scales that allow the founders of the projects to be involved in the final realization. Those days are gone if we are to be serious about entering space and exploring further. The next massive efforts in space will require that we think as a species, not as individuals, with today's decisions affecting actions 50 or 100 years in the future. Are we ready for such thoughts?

Possibly not. Instead we continually vie with each other for brief moments of glory, miniscule marks upon the fabric of history, and, in the process, threaten ourselves with extinction. Nuclear lunacy can be controlled, with a vast effort of will and trust. This trust will never be fully realized as long as prejudice and deeply held beliefs stand in the way. Only when we set the well-being of all Earth and its inhabitants as paramount will there be hope. As long as artificial gods are given control of the human endeavor, we will never shoulder the burden of responsibility for our continued survival and evolution. Until we do, we must shudder when we confront the future and hope that it will improve upon our past.

Yet these thoughts pale when we recognize that an asteroid could any day drop in and smack the planetary rump to cause unspeakable horror, or a comet may visit to bring the kiss of molecular death. That interstellar matter, seen in glorious projection against the heavens as the markings that began to intrigue Edward Barnard a century ago, contains the precursors for life has become apparent. Yet it may also contain the seeds of death. Seen from the evolutionary point of view, the death of one species always creates opportunities for others.

The exit of humankind from the galactic scene would leave no memorial. Why, then, are we not more concerned about our collective future? I would prefer that our species confront such matters with full awareness rather than rush headlong into the twenty-first century, pretending we are immortal as we seem bent upon mindless self-destruction in the face of a universe whose concerns for life on this planet are nonexistent. Why do we continue the fight among ourselves, race against race, tribe against tribe, and belief against belief, in this isolated corner of the Galaxy, as if we have all space and time in which to exist? Surely this is a sign that we have learned nothing since our brains began to ask the fundamental questions: Who are we? How did we come to be?

These questions, whose answers lie within the reach of the scientific method, are being answered; but are we, collectively, ready to confront the realities implied by the answers? It is not preordained that we should. On the contrary, it is conceivable that the grip of darkness which dogma exerts over humanity may be willing to reassert itself and that for centuries to come we may again be engaged in mindless battles over beliefs. In the meantime, however, the astronomical discoveries of recent years require that we let go of some of the more reassuring answers dreamed up millennia ago to explain our existence, early models of the universe invented when the inquiring human mind first began to function. The new views, unfortunately, do not fit too well with previous hopes and expectations, wishful thinking that rested on these ancient foundations.

And so we remain mired in struggle, and embattled in our efforts to form new ideas which may serve us better than those of yesteryear. Public consciousness struggles with new concepts, those new discoveries that rain upon us. Yet, to carry us through into the next millennium, we must consider that we are standing on a threshold, one that may bring a leap in understanding of our place in the scheme of things called the universe. Whether we succeed in unshackling our collective minds from prejudice and allow for greater exploration of space and time surely will be an interstellar matter for centuries to come.

Notes

1 Hoyle, F. (1982) *Ann. Rev. Astr. Astrophys.*, 20: 1.

2 Ibid.

3 *Star Trek.*

4 Hoyle, op. cit.

5 Hoyle, F., Wickramasinghe, N.C. (1978) *Lifecloud*, London: Dent and Son, Ltd.

6 I deliberately avoid reference to Biblical or other religious explanations which avoid the issue of whether hypotheses can be tested. The questions posed here are framed in a context which allows us to not only suggest answers, but that those answers might be tested. If answers are not immediately found, scientists are willing to live with the uncertainty that is usually our lot in any case.

7 Hoyle, op cit.

8 Hoyle and Wickramasinghe, op. cit.

9 Margulis, L., Sagan, D. (1986) *Microcosmos*, New York: Summit Books. The authors suggest that the number of species that have existed since since life began on earth may approach a billion.

CHAPTER 24

A Summary

Once Upon A Time

Edward Barnard's involvement in the search for interstellar matter began on June 29, 1892, when, as a professional astronomer at Lick Observatory, he took a photograph of the stars near the star cluster M11. He must have been thrilled by what he saw: a dark region nearly devoid of stars, "running southerly from M11 [in] a broad curving semi-vacancy."[1] However, Barnard did not elaborate on the feelings he experienced upon first seeing a dark marking looming among the stars. Such private moments are seldom reported, yet the joy they provide are rewards for the researcher exploring the universe.

His expectation that there existed vacancies between the stars was firm, however, thanks to a book he had read one night two decades earlier. During the next thirty years the idea never let go of its hold over him, although he was repeatedly tempted to change his mind about their nature. Only after his lifetime was the concept of holes in the heavens demonstrated to have been without substance.

We have traced the origin of the idea back to the mid-eighteenth century when it was first expressed by James Ferguson, although he used the concept to describe why bright nebulae appeared between the stars. It was but a small step to consider that other gaps between the stars allowed one to see empty space beyond. Ferguson wrote about the idea in a popular astronomy book that would later make a lasting impression on William Herschel, because it was the first astronomy treatise he read. That was when he was thirty-five and just beginning to turn his attention to the heavens. Later, when Herschel saw a dark marking through a telescope, he exclaimed, "Surely this is a hole in the heavens,"[2] and then proceeded to explain why such holes would exist. Star clusters, he said, would pull nearby stars toward them and leave voids in nearby space. Figure 24.1 is a modern photograph of a few of these totally dark markings and reveals how easy it must have been to think of them as true vacancies in space.

In the early nineteenth century, the Reverend Thomas Dick, another popularizer of astronomy, read Herschel's astronomical masterpiece, *On*

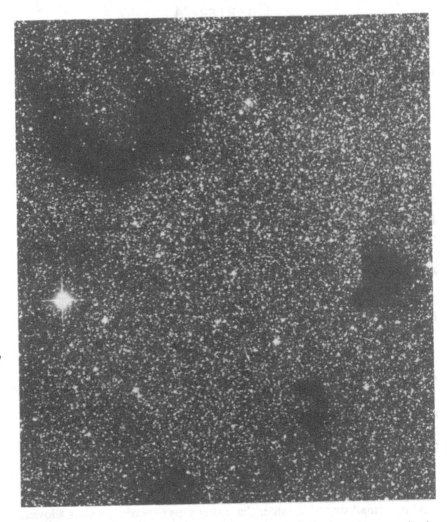

FIGURE 24.1. Dark, very dense dust clouds known as Bok globules, seen in the constellation Ophiuchus. These block out all light from distant stars and it is not difficult to imagine why William Herschel and Edward Barnard should have thought of them as "holes in the heavens" when they first saw such markings on the sky. (European Southern Observatory)

The Construction of the Heavens, in which Herschel's explanation for the vacancies was given. Dick included the description in his own treatise for the public, *The Sidereal Heavens*. In 1870, somewhere in Tennessee, a copy of this book was stolen, and the thief used it as collateral for a $2 loan from Edward Barnard, the fertile-minded Nashville teenager with a passionate, but as yet unfulfilled, interest in astronomy. Reluctant to accept the book, Barnard refused even to look at its title and after the scrounger left, young Ed, to his extraordinary surprise and delight, found that between its covers lay hidden wondrous stories of stars and constellations and instructions on how to build a telescope. The boy was elated. The book also told of dark markings in the heavens caused by vacancies between the stars.

In 1916, when Barnard was himself a world-famous astronomer and had become the pioneer photographer of dark markings in the Milky Way, he still wrestled with their nature. He wrote that "the more familiar we become with the dark objects the less appealing is the idea of openings in the sky."[3] Yet he could not tear himself away from Herschel's words and insisted that some of these regions "are undoubtedly real vacancies, or places where the stars are thinned out, but these particular ones at once proclaim their nature and leave one in little doubt as to their explanation."[4] (Unfortunately, he did not say which ones so proclaimed their nature.)

We suspect that his statement was a symptom of a lifelong struggle to change his mind. How often do we confront a parallel situation in our lives? Barnard made no error in expounding the notion that the dark regions were voids in space. He was doing no more than confront a choice between the authority he had (understandably) vested in Herschel's words, and which had so influenced his impressionable young mind, and the view he credited to A.C. Ranyard, that there existed obscuring matter between the stars. But who was he, Barnard, to make the choice and suggest that Herschel had been wrong? Furthermore, it is very difficult to change our minds about long-held beliefs, especially those espoused openly. And in his early years as a professional astronomer, Barnard had often stated that the markings were, indeed, vacancies in space.

Such is human nature! Recantation is traumatic and causes us to cling to outmoded ideas through thick and thin, no matter how strong the evidence that those ideas may be wrong. This syndrome may have something to with psychological survival! Beliefs often provide the crutches we need to get through life and, if our beliefs are deeply held or are proclaimed by great authority, the task of letting go can be frightening and virtually impossible. After all, to let go of a most cherished belief may require that you run the risk of drifting in the limbo of uncertainty until such time as another belief takes its place.

We need only look at the history of the human species to see the extraordinary power exerted by beliefs, especially those used to bolster our personal cosmologies regarding the nature of life and the universe. Religious beliefs, in particular, exert enormous power over us because they provide security in the face of the infinite unknown. Yet it is this same unknown that the scientist seeks to probe with new technologies, so as to reveal the truth beyond the comfortable illusions created by our beliefs and expectations as to the nature of things. However, scientists are human and they also struggle with issues related to changing their minds. That is why scientific paradigms,[5] are notoriously slow to change. The new generation has to wait for the previous one to die off before the paradigm (a set of widely-held scientific beliefs) is redefined.[6]

Concerning the nature of the dark markings, Barnard was always struggling on two fronts. On the one hand, a paradigm shift was occurring in astronomy and he was in the middle of it. From that vantage point it is virtually impossible to be aware of the impending change. Barnard did not know that the astronomer's world view was about to change; space that had previously been believed to be empty was about to be filled with diffuse clouds of solid and gaseous material. On the other hand, it must have been very difficult for him to deal with the personal challenge that was his, to decide between Herschel's authority (albeit of a bygone age), and the word of A.C. Ranyard (and before him, Angelo Secchi) that obscuring matter existed between the stars.

It seems odd, in retrospect, that Barnard failed to make an unambiguous choice, despite the profound insight he had "one beautiful transparent moonless night in the summer of 1913"[7] when he stood on Mt. Wilson and saw dark terrestrial clouds projected against the stars of Sagittarius. Then he understood: If physically bounded clouds of solid particles existed in deep space, they would present the same appearance when seen in silhouette against the background stars, and would look just like the dark markings revealed in his epochal photographs.

Barnard must have experienced the thrill of discovery when he made that simple, yet so profound, connection. A terrestrial cloud drifting in front of the star clouds in Sagittarius looked just like a dark marking in the Milky Way. "I have never before seen this peculiarity so strongly marked from clouds at night, because the clouds have always been too large to produce the effect."[8] It was a unique condition, small dark markings in front of the stars. Could it be that clouds of interstellar matter drifted between the stars as well? His behavior following this insight was quite consistent with the conspiracy of silence that surrounds exciting moments of insight in a scientist's career. Since these are intrinsically "unscientific" experiences, they are best not discussed. At least so goes the current dogma. And thus we have the odd situation in which our modern educational system teaches youth about the results of scientific discovery, but seldom (if ever) mentions that the fruits of the scientist's labors gave them very great personal delight and pleasure. A moment such as Barnard experienced atop Mt. Wilson must have provided him with a thrill that was a profoundly satisfying sensation which we can only infer by his nearly poetic description of the beauty of the night. To the researcher such moments make it all worthwhile.

But Barnard had to confront a practical problem as well. He was unable to *prove* his insight. Herein lay the rub. Insight is only the first step on the road called scientific progress. (By the way, this is the step that many pseudoscientists proclaim to be sufficient, which can quickly give rise to cult followers who require only the word of the master to satisfy them.) The scientist must find the tools, or the language—such as mathematics—to demonstrate the validity of his discovery or insight. In other words, the personal thrill of discovery must be turned into the beauty of shared understanding. Only then is the discovery accepted. This is the very essence of the scientific endeavor and usually requires further experimentation.

Barnard needed quantitative data to back up his insight, but such data were not available to him; nor does he seem to have wanted to pursue the search. Unequivocal proof of the existence of matter would only be manifested decades following his epiphany on Mt. Wilson. And even then, it was only after Arthur Eddington (in 1926) lent his voice of authority to the discussion that the subject of interstellar matter became "respectable." Then, very quickly, the necessary proofs for its existence were found to be at hand. By that time Barnard had, unfortunately, "shuffled off this mortal coil."

Astronomy Transformed

And so we come to the end of our journey through interstellar space, a journey that began in the minds of astronomers as they contemplated the nature of the dark markings in the Milky Way, a journey of discovery that gave many a researcher the rewards he or she sought in pursuing their careers as astronomers.[9]

Sometimes the important discoveries related to interstellar matter gave individuals great pleasure ranging from the joy of understanding to the thrill of discovery. At other times, intuitions were experienced ahead of their time, when astronomy was not yet evolved enough to follow up on those insights. Thus the likes of Secchi, Ranyard, and Barnard could not, in their lifetimes, positively demonstrate that dark markings were produced by obscuring matter between the stars.

Barnard's discovery-filled life, however, also straddled an era of transition when classical astronomy gave way to modern astrophysics. He was among the last of a breed, those great observers of the heavens, men of keen vision who spent lonely nights with eyes glued to telescope eyepieces in search of comets, planets, and nebulae, and who then described in great detail what they saw. However, their original perceptions were limited by what they could personally see through the telescope, that is, by the sensitivity of their retinas. (Barnard was a master at this and became widely recognized as having the most sensitive eyes of any astronomer in the world.)

The classical astronomer observed appearance, measured positions, and formulated classification schemes built upon brightness and shape.

Then, in the late nineteenth century, astronomers (with Barnard as a pioneer) began to *photograph* stars, comets, nebulae, and then galaxies. With the introduction of the camera and the spectroscope as new tools of the trade, an extraordinary transformation came about. The astronomer of the future would go beyond appearance to discover quantity and quality. Thus was born astrophysics.[10] The study of the spectra of stars and nebulae would allow the derivation of velocities, temperatures, densities, and the physical and chemical makeup of astronomical objects. By the end of the first quarter of the twentieth century, astronomers knew of the existence of stars, nebulae, and galaxies, and how to find their distances. Furthermore, insights into the evolutionary histories of these objects began to flower. At the same time the existence of interstellar matter was about to be established beyond a doubt.

The understanding of the nature of the "holes in the heavens" required a great deal more than portraits of their profiles. The solution was to require quantitative measurement, to a limited extent first provided by Max Wolf, as well as spectroscopic evidence, a start on which was provided by Johannes Hartmann.

Today, we know that space is filled with clouds of atoms and dust grains whose nature is partially understood from laboratory simulations. In addition, life-forming organic molecules of wondrous complexity exist in clouds within which stars are continually born, clouds spread out along the spiral arms of our galaxy and other spiral galaxies throughout the universe. What were once believed to be voids in space are, in fact, filled with diffuse matter and are permeated by magnetic fields which thread their inevitable way between the stars to control the motion of interstellar matter and the process of star birth.

The realization that space contained matter never burst upon the astronomical community as a sudden discovery, one that was immediately accepted. Barnard's often tortured struggle to come to terms with what his photographs revealed illustrated this. His dilemma highlights the fact that the acceptance of a new vision is seldom easy.

Today we must ask, what else is out there? What complex molecules lurk as yet unrecognized? What is the cause of the sudden disappearance of quasars behind interstellar ghosts moving amongst the stars? Hundreds of astronomers are engaged in the quest to solve these, and many other mysteries related to the goings on in space between the stars. Finally, there are those who dare ask, "Is there life in space between the stars?" This question, in particular, may be too far ahead of its time for us to draw meaningful conclusions from available data, yet the very question is another sign of our insatiable curiosity as we probe deeper and deeper into the intimate details of interstellar matters. It is left to future generations to explore farther and to answer these questions.

Notes

1 Barnard, E.E. (1892) "Photographs of the Milky Way." *Astrophys. J.*, 1: 10.

2 Hoskin, M.A. (1964) *William Herschel and the construction of the heavens*, New York: W.W. Horton and Co., Inc.

3 Barnard, E.E. (1916) "Some of the dark markings on the sky and what they suggest." *Astrophys. J.*, 43: 1.

4 Ibid.

5 Kuhn, T. (1970) *The Structure of Scientific Revolutions* Chicago: University of Chicago Press.

6 Ibid.

7 Barnard, op cit.

8 Ibid.

9 However, we have not covered every facet of the subject, and never pretended we would. Many aspects of interstellar matter as a modern science have been left out of our essays. We have paid scant attention to the fundamental cycles by which interstellar matter, gaseous and solid, is continually cycled and recycled through stars and back into space again. The injection into space occurs in many forms of star death, from planetary nebulae and novae to the destruction of entire stars in violent supernova explosions. All of these feed processed matter into space and it is then incorporated in future generations of star and planet formation.

10 Osterbrock, D.E. (1984) *James Keeler: Pioneer American astrophysicist.* Cambridge: Cambridge University Press. This book presents a delightful record of this era of transition.

Index